Holography Marketplace
1st Edition

The original version of this book contained holograms from the vendors and many of them are no longer in business. Therefore this version of the book contains everything that was in the original version but it has no holograms.

Edited by Franz Ross and Elizabeth Yerkes

PREFACE

The Holography Marketplace (HMP) is an annual publication and is intended to provide both an overview of the holography industry and a database of businesses in the industry.

In this first issue of the HMP we cover the field of Artistic Holography. There are other fields of holography such as the making of Holographic Optical Elements (HOE) and the use of holography in NonDestructive Testing (NDT). Both NDT and HOE will receive a thorough presentation in coming editions.

In compiling this first edition some editorial license had to be exercised on what businesses were listed. We decided to list all businesses that have been in operation for 1 year or more. If you are a new business and were not included in this edition, you will be listed next year.

We recognize that we are not infallible in our knowledge of the industry and we do not claim to know everyone. If you are a business that is not listed, please write or FAX us on your business letterhead. Include information about your business so we can list you appropriately in our next edition. If you feel we have omitted anything or made any factual errors, please let us know. We do read the mail and will make every effort to improve and expand in subsequent editions.

Franz Ross   Elizabeth Yerkes

Ross Books

P.O. Box 4340

Berkeley, Calif.

USA,94704

Tel.l-415-841-2474

FAX: 1-415-841-2695

CIP Information

Holography Marketplace.

Bibliography: p.
Includes index.
1. Holography.
2. Holography industry–
Directories.
TA1542.H649 1989 338.4'76213675
88-31798

ISBN 978-0-89496-094-9

TABLE OF CONTENTS

INTRODUCTION

HOLOGRAPHY MARKETPLACE 1989

There are a tremendous number of options available in 1989 to the individual who wants a hologram made. More studios are capable of larger formats, wider color ranges and combinations, and more ambitious subject matter than ever before. The options in lighting and presentation have also been broadened, enabling holograms to be displayed in new and exciting environments. Massmanufacturing techniques continue to be refined for each major type of hologram, meeting the demand for increased quality and manufacturing runs.

If you are new to the holography industry, this introduction is for you. Before going to the names and addresses sections, it would be good to understand some basic terminology and become familiar with the fundamental procedures used in holography. In this way you can carry on an intelligent conversation with your supplier and avoid wasting time.

It is also wise to see samples of a variety of holograms. Although this book describes different holograms and their best qualities, there is no substitute for seeing holograms first hand. If you do not have a gallery or place near you where you can see holograms, there are a number of mailorder businesses listed in this publication that would be happy to send you a catalogue so you can order samples by mail. Holograms are unlike anything else and there are many types with distinct and striking variations.

In this introductory chapter we are going to define some fundamental terminology. If you take time to understand the introduction you will be well equipped to read the rest of the chapters that describe in detail what the different types of holograms can do for you and what kind of design and subject matter works well with them.

What is a Hologram?

We want to discuss holograms from a user's point of view. How are they made? What are the different types? What are the good and bad points of each type?

A hologram is a three-dimensional picture that is made on a photosensitive plate or film using a laser as the light source. The advantage holography has over photography is that it can have full parallax, while photography has none. By this we mean that the viewer, while looking directly at the plate and viewing the hologram, can tilt the plate slightly to the left or right and see around the image. The amount of depth the hologram has depends on how it was made. With a photograph, of course, you see only the flat image regardless of how you move the print. Over the years there have been a number of items marketed that are optical illusions and try to fool the brain into thinking it is seeing a threedimensional image, but holograms actually are three-dimensional images.

Perhaps the best way to familiarize you with some of the basic terms used in holography is to describe making a simple hologram which we will do on the next page.

Making a simple hologram

We are in a holography studio and we see a table. On that table is a laser, some mirrors and a photosensitive plate in a plate holder. They are arranged as below:

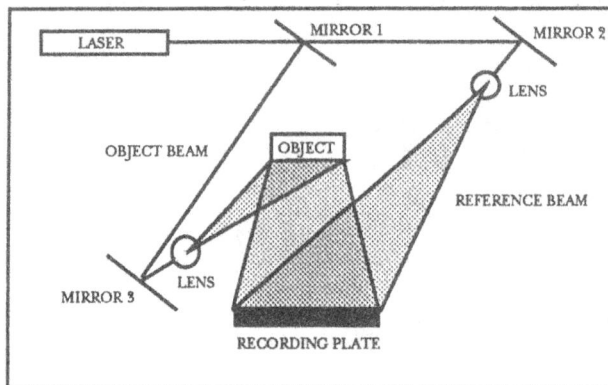

This is a simple set-up. Let's say we are ready to make our hologram exposure. We turn on the laser. The laser beam strikes mirror 1 which is a partially mirrorized mirror. Because it is only partially mirrorized, part of the beam goes through mirror 1 and part of it is reflected. Mirror 1 is referred to in the trade as a "beamsplitter" for this reason.

The beam that goes through mirror 1 then hits mirror 2. Mirror 2 reflects the beam toward the photosensitive plate. A lens is placed in the path of the beam after striking mirror 2 so the beam will spread out and cover all of the plate. The beam that went through mirror 1 and struck mirror 2 is called the "reference beam" because it never strikes the object being holographed on its way to the photosensitive recording plate.

The other beam reflects off mirror 1 and strikes mirror 3 which aims the beam toward the object. A lens is placed in the path after mirror 3 to spread the beam out so it illuminates the entire object. The light is then reflected off the object and strikes the recording plate. This beam is called the "object beam".

Providing the length that the object beam travels on its way to the recording plate is equal to the length that the reference beam travels on its way to the recording plate, you will get a hologram when the plate is exposed for the proper amount of time and developed.

VIEWING A FINISHED HOLOGRAM IN LASERLIGHT

Our exposure is done. Next we take the exposed recording plate out of the plate holder and develop it. We now have a finished hologram. Inspecting the hologram plate, we see that the plate is transparent except for some patterns made by the object and reference beams.

To view the finished hologram, we do the following:

1) Put the hologram plate back in the plate holder.
2) Take mirror 3 and the object off the table.

With the laser light turned on, the only thing now illuminating the plate is the reference beam. Since the plate is reasonably transparent, the laser light will shine through the plate to us on the other side. Looking into the plate, you will be able to see your object just as though it were really there and at the same depth and position it was when originally shot. Moving from side to side while looking through the plate, you can even see around the image.

A detailed explanation of why this happens would be exhaustive. A simple explanation might go like this: the object and reference beams strike the photosensitive emulsion at the same time and create patterns in the photosensitive emulsion referred to as interference patterns. After development, we aim only the reference beam at the plate and at exactly the same angle that it originally exposed the plate. As the reference beam goes through the plate to us on the other side, it bends as it strikes the interference pattern. The interference pattern, which was created by light from the original object, bends the light passing through it so the light focuses back in space to recreate the image of the original object at its original position in space. Thus, the light recreates a threed mensional image of the original object.

Note that when making the hologram we used a laser so we could get a single beam of light at one wavelength. Also note that the object beam and reference beam had to travel the same distance so the light waves were right in step with each other. This way, the light waves meeting on any part of the plate cancel each other out if they are out of phase and reinforce each other if they are in phase. Had we made the exposure using regular light, with many different wavelengths, the wavelengths would combine to form many interference patterns and thus do nothing more than fog the plate.

In viewing a developed hologram plate, it is important that the reference beam shines on the hologram plate at exactly the same angle that it originally exposed the plate in order to clearly see the image of the subject in its original position.

VIEWING A FINISHED HOLOGRAM IN WHITE LIGHT

Although it is necessary to use a laser to make the hologram, it is not always necessary to use a laser to see a hologram. In fact, most holograms can be seen in sunlight. Let's say that you have made one of these sunlight-viewable holograms (also called white light-viewable holograms). Once you have made the hologram it recreates the original object's image when the wavelength that was used to make the hologram passes through it. Sunlight consists of many wavelengths and includes the wavelength at which the hologram was originally exposed. If you shine sunlight through the hologram, the original wavelength that exposed your hologram passes through the interference pattern and recreates the image.

There are problems here because all the other wavelengths in sunlight are also coming through and being scattered around. This unwanted light tends to compete with the image your hologram is trying to form in space and can wash it out. Consequently, the more difficult your image is to form (i.e. the more depth or projection your image has), the more you want a light source with very few wavelengths that might compete with the wavelength at which the laser made the exposure.

Thus you will find that holograms made to be viewed in a wide range of lighting will use subjects that have shallow depth. When you go into a shop that sells holograms, you will find that the shop generally has subdued, overhead lighting with "pointsource" lights focused on the holograms. A pointsource light is a clear light bulb with a single filament. This serves the dual purpose of creating a lighting environment that is very pleasant to be around as well as reducing the number of wavelengths competing with each other so that holograms with much more depth and projection can be displayed. Clear light bulbs with single filaments are available at any store with a large selection of light bulbs.

VIEWING A MULTI-CHANNEL HOLOGRAM

This is an interesting double exposure (or multiple exposure) effect. It is possible in holography to make a first exposure just as we described in the earlier diagram and then take a second object to be holographed, graphed, put It III the pOSItion of the first object, change the angle of the reference beam and make a second exposure on the same plate. You then develop your hologram.

Holding the finished hologram up to a light source (which acts as your reference beam), you see the first object when the angle between the light source and the hologram plate is the same as the angle between the first exposure's reference beam and the hologram plate. Turning the finished hologram, you see the second object when the angle between the hologram plate and the light source is the same as the angle was between the second reference beam and the hologram plate.

In other words, you can tilt the hologram back and forth and as you do so different objects appear. This is what is referred to as a multi-channel hologram because there is more than one channel or angle between the hologram plate and the reference beam that plays back images. This is very common in embossed holograms. Many of them will play back more than one scene when tilted back and forth. It can be used to show before and after sequences or whatever a designer can dream up.

Transmission and reflection holograms

What we just discussed was a hologram that was viewed by letting light go through the plate to our eyes. That is called a transmission hologram because the light passes through the plate to us. It is also possible to make a hologram where the light reflects off the surface and back to our eyes for us to see the image. This is called a reflection hologram. Let's look at the types of transmission holograms that exist and then discuss the reflection hologram.

TRANSMISSION HOLOGRAMS

There are two types of transmission holograms. One type of transmission hologram can only be viewed by using a laser light to illuminate it. The second, and much more popular type can be seen in sunlight or under any point-source light These sunlightviewable holograms are also called white lightviewable holograms. One of the popular variations of the white light hologram is the rainbow hologram; so called because it generally displays rainbow-like bands of color as you shift your point of view up and down while looking at it. As with any trade, one of the problems in holography is

R I C H A R D   B R U C K   H O L O G R A P H Y
3 3 1 2   W.   B E L L E   P L A I N E   C H I C A G O   IL   6 0 6 1 8

3 1 2 . 2 6 7 . 9 2 8 8

jargon. Several names are sometimes used to describe the same item.

REFLECTION HOLOGRAMS

Now we will discuss how reflection holograms are made. Look at the set-up we diagrammed. If we transfer the reference beam around with mirrors so it illuminates the recording plate from the back instead of from the same side as the object beam, we create a reflection hologram. All reflection holograms are white light-viewable and, as their name implies, all reflection holograms are illuminated from the same side as the viewer (so the light comes from your side of the plate, strikes the plate and reflects back to your eyes).

SUMMARY

So far we have seen that there are three types of holograms:

1) *Laser-viewable transmission holograms* (viewable under laser light only)

2) *White light-viewable transmission holograms* (Viewable under any point-source light, also called sunlight-viewable holograms)

3) *Reflection holograms* (all reflection holograms are sunlight-viewable) Every hologram that is made is in one of the above three categories and adheres to the basic principles described above. The set-ups, however, can be wildly elaborate with mirrors and gizmos all over the place for special lighting and effects. This is a very open-ended and creative art form.

Emulsions

We know that holograms are either transmission or reflection depending on what exposure setup we use, but what about the photosensitive emulsion that actually records the hologram pattern? How many emulsions are there and what are they? On what are they coated?

Answering the last question first, the photosensitive emulsions are applied to either glass or film. Both are used in the industry. Film is obviously less expensive and easier to mass-manufacture but glass is more popular for original masters.

There are four main types of emulsions.

1) Silver halide
2) Dichromate gelatin-also known as D.C.G.
3) Photopolymer
4) Photoresist

SILVER HALIDE

Both transmission and reflection holograms are made using silver halide as an emulsion. Almost all masters made are silver halide laser transmission holograms.

There is a big market in silver halide reflection holograms as well. They are very common in shops that sell holograms and most of them exhibit good depth and projection.

DICHROMATE AND PHOTOPOLYMER

Holograms made using these two emulsions look very similar. Both of these emulsions are viewed like reflection holograms. Not much has been seen of photopolymer products but we expect to see a lot in the near future. The advantage of using dichromate or photopolymer emulsions for your hologram instead of embossing your hologram is that DCC and photopolymer have much more clarity. This is because they have a non-mirrorized backing and therefore do not reflect background light which competes with the image. We will explain this in detail in the next chapter.

PHOTORESIST FOR EMBOSSED OR HOT-STAMPED HOLOGRAMS

This emulsion is used exclusively in embossed holography. Photoresist is a photosensitive emulsion that you can use to record your hologram on just like the other three emulsions. It is then used as a master in the process of making a die for stamping out embossed holograms. A detailed explanation of this will be given in chapter one.

Embossed holograms are used very widely. They are on your bank cards, they are on magazine covers, food containers, and many other products. Their primary uses are for identification (security) and for point-of-purchase attention getters. Their advantage is that they can be mass-manufactured and machineapplied to products by the millions. We may see competition in the future from DCC and photopolymer holograms, as they are clearer, but the ability to apply large quantities of holograms at a comparatively low unit cost favors the embossed hologram at present.

Lasers

We will not go into depth discussing lasers in this book but the difference between CW and pulse lasers should be spoken about now. There are two major kinds of lasers; the Continuous Wave (CW) laser and the Pulse laser. The CW laser emits a continuous wave of laser light whereas the pulse laser emits laser light in bursts.

CONTINUOUS WAVE LASER

The power of a CW laser is measured in watts (w) of power. The CW laser is by far the most common laser used in holography. In holography labs, most of these lasers fall in the 5 to 50-mw (milliwatt) range. One of the great problems with CW lasers is that they cannot make the extremely short exposures necessary to capture a live subject. Consequently, there must be absolutely no motion at all during the exposure with a CW laser. The exposures with CW lasers can take a fraction of a second to several seconds. Because

there cannot be any motion at all during our exposure, we need to get rid of the vibration coming from the ground. To do this we make or buy a vibration isolation table on which to put our laser, optics and objects. Since it is absolutely critical that we have no motion at all, the subjects that we holograph with CW lasers have to be "dead" objects. Feathers might move in the breeze and living things move too much.

Remember that what we are recording on the plate is the reference beam and the object beam converging (or interfering) with each other at the plate. If the object moves fven a microscopic amount (on the order of wavelengths) from one moment to the next, we will be recording two different interference patterns and the hologram will fuzz out or not even be there (just like the photographic daguerreotypes of the 1800's only much less forgiving).

PULSE LASER

Pulse lasers, on the other hand, emit extremely quick bursts of very powerful laser light. Consequently, the exposure time is much shorter than a CW laser. Exposures are in picoseconds. One picosecond is one trillionth of a second. You don't need a vibration isolation table for the pulse laser. What can you shoot? Anything you want. You can shoot an entire room of people belly dancing in costumes of paper with feathers in their hair, and birds flying around the room. Why such freedom? Because your subject can't move significantly in a picosecond.

What are the drawbacks of pulse lasers? Why doesn't everyone buy one? The answer is money. They cost in the US$60,000. range and require a lot of extra overhead and care. Lasers don't last forever and when a pulse laser bums out it is expensive to fix. Hologrphers are anxiously awaiting, with cash in hand, a low-cost, easily maintained pulse laser.

The reason this discussion on lasers is important is that who you go to for your hologram depends on what you want holographed. If you have a corporate logo or some other object that does not move, a CW laser can do just as good a job as a pulse laser. There is no need to pay extra to have your hologram made by pulse laser if you do not need it.

The H-1 master

It is important that we cover the topic of the H-1 master in the introduction because it is a fundamental procedure in almost every job. H-1 stands for "hologram one" because it is the first master you make on the path to your desired final hologram. Sometimes there is more than one master that needs to be made. If so, the next master is called the H-2, and so forth.

One of the big problems that holographers used to have was placing their object to be holographed exactly where they wanted it. Suppose, for example, that in the set-up we diagrammed earlier in this chapter, we wanted the object in our final hologram to appear half in front and half behind the recording plate. How would you do it? This problem was solved by the following procedure:

1) You go ahead and make a laser-viewable transmission hologram just the way the set-up is diagrammed. We call this H-I because it is our first hologram master.

2) Since the H-I hologram creates an image of the object, why not use the image made by our H-I as our subject and make a hologram of the image. In other words, make a hologram of our hologram.

3) It sounds strange, because you are making a hologram of an image and not an object. But it works. Now, since you can make a hologram of H-I 's image, take time to move the image around to wherever you want it positioned. Adjust your recording plate so that the image of the object is half in front and half behind the plate. The problem is solved.

In summary, here are three good reasons for making an H-I:

1) H-I allows you to reposition the image of your subject. When you reposition your image from H-I, you can make your subject focus out in front of the recording plate, behind the plate or anywhere within the limits of your equipment (You are usually limited by the laser's coherence length and the quality of the optics). The creative potential here is enormous because you are able to move solid objects around like they are ghosts. You can have two objects occupying the same space, etc.

2) It gives the holographer a chance to brighten up the image. Since you can move your image anywhere, you can focus the image right at the recording plate. This concentrates the light directly on the emulsion and brightens up the image considerably. This is commonly done for silver halide reflection holograms.

3) Another great advantage to making an H-I is that if you don't like the position of your subject astride

the recording plate, you don't have to find the original subject and set it up again. This can be important if you were shooting the belly dancers we discussed earlier.

The H-I master is used in the process of creating most holograms. It is techinally possible to get to some desired holograms by skipping the H-I but the results are generally very inferior. H-I is almost always used for commercial jobs of value.

Color in Holography

COLOR VARIATIONS
Reflection holograms are usually monochromatic under white light. However, it is possible to change their color. One way this can be done is by swelling the photosensitive emulsion using a specific percen tage of an inert chemical, and changing the reference angle a certain degree (to correspond with the color you want) and then making the exposure.

Let us say you want to have two subjects in the hologram.You want each one to be a different color and you want them seen by the viewer at the same time (ie. not multi-channel). For two colors, the first exposure is made, the subjects are exchanged, the emulsion is preswelled to another color's percentage, the reference angle is changed and the second exposure is made.

The pre-swelling makes the whole recording surface fatter for the exposure. When the hologram is processed the swelling agent is washed out and the thickness of the hologram returns to normal. At normal thickness, the hologram selects not the red wavelengths by which it was made, but instead selects shorter wavelengths-ranging from an orangered to violet, depending on how much it was preswelled.

For registration purposes the reference angles also must be changed with each pre-swelled color exposure. If you make a two-color hologram by preswelling the emulsion and not changing the reference angle accordingly, the result will be a twochannel hologram. Looking straight at the plate you see a red apple, and turning the plate a little the apple disappears and blue grapes appear.

There are many complications with the above procedure: swelling agent ratios, changing of reference angles, and multiple exposures. Added to this is the problem that the hologram cannot be copied without going through the whole process again (it cannot be mass-manufactured). Consequently, this method is used for a single hologram or for small-run, custom work.

FULL COLOR HOLOGRAPHY

This is not easy to do (or everybody would be doing it) but it is easy to understand. In the back of the human eye are light sensors called rods and cones. They are acutely sensitive to three colors: red, green, and blue. The human brain receives only this information and uses it to create our full color world. The idea, then, is to make a full color hologram by using three colors of laser light (red, green, & blue) instead of one as the exposing light source.

Unfortunately, it is not anywhere near as easy as it sounds. Only a few holograms have been made that look close to "true color". There is, however, considerable research going on. Paul Hubel of the Department of Engineering Science at Oxford University, England recently produced a reasonably good true color hologram after two years of research.

Another method, beautifully executed in the hologram "DOJO" by Toshihira Kubota of Kyoto Institute of Technology, is to layer multiple emulsions of the same subject on top of each other.

Commercial true color holography is coming but it is still in the research stage.

Overview Of holography

On the next page is a chart showing the most common artistic holograms which we have briefly introduced here.

In the next chapter we will discuss the holographic stereogram and other holograms in more detail.

art
design
commercial

limited
custom
& mass
production
& consultation

...

graphi_{e.}
édition

edition
co-production
consultation

...

. import

gifts & cards
jewellery
decoration
collection
art works

...

holography
distribution
& retail

display systems

exhibitions

...

IBOU inc

C.P. 214
cap-de-la-madeleine
québec, canada
G8T 7W2

* TM of IBOU inc.

Flowchart: Holography production processes

SUBJECT IS HOLOGRAPHED FROM
- FLAT ART • MODEL • LIVE SUBJECT WITH A PULSE LASER • A SERIES OF PHOTOGRAPHS, OR MOTION PICTURES, OR COMPUTER-GENERATED IMAGES TO MAKE: *

A DICHROMATE GELATIN, WHITE LIGHT-VIEWABLE REFLECTION HOLOGRAM
- H-1 MASTER(S) → H-2 NOT USUALLY USED → FINAL PRODUCT: REFLECTION HOLOGRAM ON DICHROMATE GELATIN EMULSION: USUALLY SEALED IN GLASS

A PHOTOPOLYMER, WHITE LIGHT-VIEWABLE REFLECTION HOLOGRAM
- H-1 MASTER(S) → H-2 NOT USUALLY USED IN PHOTOPOLYMER PROCESS → FINAL PRODUCT: REFLECTION HOLOGRAM ON PHOTOPOLYMER USUALLY SEALED IN PLASTIC

A WHITE LIGHT-VIEWABLE REFLECTION HOLOGRAM
- H-1 MASTER(S) → H-2 (OPTIONAL MASTER(S)) → FINAL PRODUCT: REFLECTION HOLOGRAM, VIEWABLE IN WHITE LIGHT. ONE COLOR OR SINGLE COLORS COMBINED

A SILVER HALIDE WHITE LIGHT-VIEWABLE TRANSMISSION HOLOGRAM
- H-1 MASTER(S) → H-2 (OPTIONAL) MASTER (S) → FINAL PRODUCT: WHITE LIGHT-VIEWABLE TRANSMISSION HOLOGRAM ON SILVER HALIDE EMULSION

A SILVER HALIDE LASER-VIEWABLE TRANSMISSION HOLOGRAM
- H-1 MASTER(S) → H-2 (OPTIONAL) → FINAL PRODUCT: LASER-VIEWABLE TRANSMISSION HOLOGRAM ON SILVER HALIDE

A HOLOGRAPHIC STEREOGRAM
- H-1 MASTER(S) → H-2 (OPTIONAL MASTER(S) NOT USUALLY USED IN HOLOGRAPHIC STEREOGRAM PROCESS) → FINAL PRODUCT: WHITE LIGHT TRANSMISSION HOLOGRAM, CAN BE PUT ON ANY EMULSION USUALLY MOUNTED ON CURVED OR CYLINDRICAL DISPLAY (CROSS STEREOGRAM). CAN BE SHOT DIRECTLY FROM PHOTOS, MOTION PICTURES, OR COMPUTER-GENERATED IMAGES

AN EMBOSSED HOLOGRAM
- H-1 MASTER(S) → H-2 (OPTIONAL MASTER(S) NOT USUALLY USED IN EMBOSSED PROCESS) → MASTER OR SUBJECT IS HOLOGRAPHED ONTO PHOTORESIST AS A WHITE LIGHT TRANSMISSION HOLOGRAM → PHOTORESIST IS ELECTRO-PLATED WITH SILVER-NICKEL IN CHEMICAL BATH. METAL SHIMS ARE THEN MADE FROM METAL MOLD FOR EMBOSSING PRESS → FINAL PRODUCT: EMBOSSED HOLOGRAM ON MIRROR-BACKED FILM CAN BE MADE INTO HEAT-RELEASE FOIL (HOT FOIL) OR INTO EMBOSSED STICKERS

LEGEND
- ——— = USUAL PROGRESSION OF STEPS
- ······ = CAN BE DONE BUT IS NOT TYPICAL ROUTE
- ∗ = ALL H-1 MASTERS ARE LASER-VIEWABLE TRANSMISSION HOLOGRAMS

CHAPTER 1

TYPES OF HOLOGRAMS

In this chapter we will discuss each of the following types of holograms.

Dichromate
Photopolymer
Silver Halide Reflection
Silver Halide Transmission
Embossed
Holographic Stereogram

The creation of artwork and subject matter to be holographed will be covered in chapter two. In this chapter, we will discuss the good and bad properties of the holograms listed above and, where needed, the various steps used to make the hologram. We assume that you have read the introduction and we will now make use of the terms we defined there. If you are not familiar with a term, there is a glossary and an index to consult.

Dichromate and Photopolymer

The dichromate hologram and the photopolymer hologram are very similar to each other in appearance, application, size and function to the untrained eye. They are both viewed like reflection holograms and the dichromate hologram (also known as D.C.C., or dichromated gelatin) is the most wellknown of the two. The DCC is usually seen on a sealed glass substrate. The DCC's images are usually close to the hologram surface and display the image in almost any available light. The most common DCC sizes are from small, inch-and-a-half circles, up to standard sizes of 4"x 5". The DCC holds metallic colors very well. Images of the inside of a watch look almost identical to the original watch. Photopolymer holograms are made by Polaroid and are relatively

new. There are very few examples of photopolymer on the market but we expect to see a lot more in the near future.

MANUFACTURING
It is not cost-effective to do an extremely short run of photopolymer holograms but the price is competitive with dichromate and silver halide in long runs.

DCG can be made in very short runs. Even orders of 1 to 100 are done commercially. The way short runs are done is by shooting the final DCG hologram directly from the object and thus skipping any mastering. This shortcut is used commercially because it allows businesses to test-market their product in short runs without the expensive mastering costs involved in a mass-manufacturing run.

Probably the biggest advantage both DCC and photopolymer holograms have over embossed holograms is their clarity. This is due to the fact that DCC and photopolymer products do not have a mirror backing like embossed holograms. The mirror backing tends to flood the hologram with unwanted light and wash out the image.

Silver Halide Reflection

Silver halide reflection holograms are very common in most shops that sell h olograms. Three reasons they are popular are:

PROJECTION
The most popular holograms on the market today are either Embossed, Dichromate, or Silver halide reflection holograms. Transmission holograms are popular among artists but are difficult to display be

cause they require lighting from behind. Consequently, their sales volume is low.

Most of the holograms made in the embossed and dichromate format are made with shallow depth and projection so that they display well under most lighting conditions. This allows them a very wide market. The silver halide holograms, on the other hand, are generally made with much more depth and projection. They expect to be sold in shops where they can be illuminated by a clear light source. Consequently, when you look at what is generally available for sale, silver halide holograms tend to have much more projection and depth than DCC or embossed holograms.

There are notable exceptions to this and the trend may change. Common examples of silver halide reflection holograms are 4 x 5 inch holograms that project images two to three inches in front of the plate. They come in standard sizes up to 12x16 inches and the image can project or recede up to a foot in front or behind the plate under optimal lighting conditions. The creation of silver halide reflection holograms is limited mostly by the size of optics and the power of the laser needed to reduce the exposure time for larger shots.

DISPLAY

Another good thing about silver halide reflection holograms is how they are displayed. Since these are reflection holograms, they are viewed with the light source on the same side as the viewer. Therefore, they can be hung on the wall and treated like a painting. A nice clear light source or spotlight is usually arranged from overhead or sometimes the hologram is sold with its own light source and display stand. DCC and embossed holograms are also viewed as reflection holograms and can be treated the same way.

MANUFACTURING

Machines that mass-manufacture copies of reflection holograms are capable of runs in the tens of thousands. The mass-manufacturing quality, which at first was poor, has improved considerably.

Silver Halide Transmission

Silver halide transmission holograms come as:

1) laser-viewable transmission holograms.

2) white light-viewable transmission holograms.

Let's discuss each of these as they apply to artistic holography.

LASER-VIEWABLE TRANSMISSION HOLOGRAMS

Although sometimes displayed and sold for its artistic merit, the laser-viewable transmission hologram is almost exclusively used for the H-l or master hologram. Laser-viewable transmission holograms demonstrate amazing depth and projection when the right equipment is used. It should be noted that the depth of the hologram is not so much a function of the power of the laser as it is the coherence length of laser light You can read more about the coherence length from several of the books listed in the bibliography.

Theoretically, the maXImum image projection in front of the hologram can be as great as the depth. Unfortunately, it is more difficult for our brains to make sense of projected images. Because of this and the fact that there are some optical distortions in the image planing process, projected distances in transmission holography are usually kept under four feet. It should be pointed out that projected hologram images generate one of the highest "shock and thrill" responses from viewers. A fair number of first time viewers respond by waving their hands through projected images in disbelief.

As we mentioned, the laser-viewable transmission hologram is most often used as a master hologram (H-l) . Transfer copies (simply making another hologram using the image on H-l as the subject) are then made from the H-l. These transfer holograms can either be other laser-viewable transmission holograms, white light-viewable transmission holograms or reflection holograms (always white lightviewable).

Laser transmission holograms have the widest parallax (the ability to see around an object side to side) and resolve the greatest depth of objects. There are laser transmission holograms, for example, of people and objects in a 4000 cubic foot room exposed with a pulse laser.

Pseudoscopic and Orthoscopic Images

Both laser-viewable and white light transmission holograms share an interesting property. In the introduction we described making a laser-viewable hologram. After developing the hologram, we put it back in the plateholder and, with no object on the table, we were able to see the original object in its original position on the table.

Now comes the interesting feature. Take the transmission hologram out of its plateholder, flip it over, and put it back in the plateholder. Step back and look at the plate. You will see your image forming out in front of the plateholder (between you and the plateholder). It focuses in air the same distance in front of the plateholder as it originally sat behind the plateholder. You will also note that the image is a pseudoscopic image. What is pseudoscopic? An image as we normally see it in everyday life is an orthoscopic image. A pseudoscopic image is the exact opposite. What was top is now bottom, what was right is now left, writing reads backwards, etc.

A pseudoscopic image is an exciting effect, but it can be confusing to the viewer. Some artists, however, have produced exciting pseudoscopic work using geometric shapes like wide spirals, pyramids and cones. One example is a hologram where the object behind the hologram plate starts as a narrow spiral and spirals wider into the background. Flip the hologram over and the image is reversed, inside out, backwards and upside down. From this flip-side, the spiral's image displays the widest spiral circumference closest to the viewer and the tighter spirals away from the viewer.

WHITE LIGHT TRANSMISSION HOLOGRAMS
A white light transmission hologram, as discussed in the introduction, is a transmission hologram that can be seen in sunlight. There are several types of white light transmission holograms. Two types that we will discuss are achromatic and rainbow transmission holograms.

Achromatic Transmission Holograms
Achromatic transmission holograms are black and white transmission holograms. The achromats are reminiscent of nineteenth century daguerreotypes. For those who perhaps see the rainbow colors as too brash and want a more serious look to their hologram subject the achromat is a handsome choice, though restricted in viewing angles.

Rainbow Transmission Holograms
One of the problems with the achromatic transmission hologram is that it has very little depth. This problem was solved with the rainbow transmission hologram. The rainbow hologram is made by taking a laser-viewable transmission hologram master (H-1) and making a transfer hologram of it through a mask that has a narrow slit in it. The resulting rainbow transmission hologram has no vertical parallax (you can't see over the top of the image) but has a wide horizontal parallax. Rainbow holograms get their name from the fact that as you shift your view-

ing angle up and down the image appears bathed in rainbow-like color changes.

Some rainbow transmission holograms are displayed in art galleries on glass plates and on film. However, they are much more popular in two other formats. The two most popular forms in which you see rainbow transmission holograms are as:

1) *embossed holograms.* In an embossed hologram, the light goes through a rainbow transmission hologram that has been embossed in plastic, strikes a mirror backing and reflects back through the transmission hologram to your eyes.

2) *holographic stereograms.* If you have seen "moving holograms" you most likely have seen one of these. This is one of the most exciting fields of holography. Included in this category are Cross holograms, computer- generated holograms and the recent Holodisk® manufactured by Advanced Dimensional Displays, Van Nuys CA.

We will now discuss how each of these is made.

Embossed Holograms

Embossed holograms are created in several steps. The most common way they are made is as follows:
1) The H-1 master is made.
2) A rainbow transmission hologram is made on photoresist (photosensitive emulsion) from H-l.
3) The photoresist is etched to relieve the hologram patterns.
4) The photoresist is plated with silver and nickel (photoresist now behaves like a metal mold).
5) The metal mold, or shim is removed from the photoresist. It now has holographic patterns on it
6) This shim is used as a stamping die and stamps the holographic patterns into plastic.
7) This plastic with the holographic pattern stamped in it has a mirror-like backing so light comes through the plastic, strikes the mirror-like backing and, reflecting back out, displays the white light rainbow transmission hologram.

Let's look at these steps in a little more detail:

Steps 1-2: There is a possible shortcut here. Sometimes flat art can be directly exposed onto the photoresist material. This is recommended for simple projects only.

Photoresist is a very tricky medium to record on holographically. It is nowhere near silver halide in responsiveness and requires long exposures even with lasers of several watts of power. A small holography studio would have to be equipped with an expensive laser and heat-resistant optics. New alternatives to this medium of photoresist are being researched.

The typical turn-around time is four to six weeks to receive excellent photoresist plates that are fit for metallizing. The holographer should make several photoresist plates of good quality to cover any problems that might occur in the metallizing phase.

Steps 3-4: After the photoresist hologram IS checked for clarity, brightness and overall quality it goes on to the metallizing stage of production. A thin layer of silver is deposited on the photoresist. Silver by itself cannot withstand the stamping pressure, so additional coats of a nickel-based material are deposited to reinforce the back of the silver. When it achieves the desired thickness, the nickelsilver shim is pulled from the photoresist plate and this becomes a hard, stamping die.

Steps 5-7: The first shim must be perfect. This first shim can have several shims made from it and in turn several shims made from those. The heat and pressure of embossing several thousand to several million holograms will wear out the shims so extras should be made. A.ly wear will be very obvious if the shims are not changed regularly.

Producing the final hologram is the job of the embosser. The embossed holograms are made by stamping the shim onto a heated polyester material which has a metallized backing. Although not used widely, colored metallized backing is an option. Most often the silver color is selected.

APPLYING EMBOSSED HOLOGRAMS

The press embosses the holograms on rolls of stock which, in the most common commer cial work, have an image area of six square inches for each impression. The roll consists of a continuous ribbon of six-by-six inch embossed squares separat ed from one another by about one-half inch. The holograms are often arranged on each six-by-six inch square as nine two-inch squares of different holograms, but you can use the entire 6x6 square for one hologram.

There are a number of options for displaying your final embossed product. Embossed holograms can be produced as stickers cut to your size specifications. Typical of the final holograms are individual peeland-stick holograms. If the stickers are to be placed onto some surface be sure the substrate the sticker is on is thick and strong enough not to conform to a textured surface onto which it might be applied. The best surface on which to stick a hologram is smooth and rigid so you can be sure that the hologram is flat and consequently able to reconstruct its image properly. There are applicator machines to apply the sticker to your product, but it is often done by hand.

Instead of stickers, you may apply the holographic foil directly to a surface like a book cover. If you choose this route, be absolutely certain of the strength and smoothness of the cover material. The embossing foil is very thin-like the foil around individual sticks of chewing gum. If the hologram has ripples in it as a result of a bumpy surface, the image itself might appear rippled.

It is a good idea to have the surface approved by your holographer and the printer, together, before going into production. Also beware of coated paper stock and printed surfaces. These surfaces could present a problem for adhesion of holograms.

Hot-stamping the hologram onto a paper surface is another popular way to apply your hologram. The "hot-foil" application is a heat-release process and keeps precise registration better than methods of sticker application. Although this process appears more expensive, when you consider the costs of a very large run using sticker application they both can come out about equal in cost Smaller runs seem to favor sticker application.

Regardless of which method you choose, it is advisable to be there when when the holograms are applied so you can check the quality.

EMBOSSING PULSE HOLOGRAMS AND HOLOGRAPHIC STEREOGRAMS

Pulse hologram images have just recently been used in embossing and so far have been very successful. Be sure the pulse holographer knows that you want this to be an embossed hologram. It is a more complicated approach which may include a step to reduce the pulse hologram to the six-inch square size of the embossing machine, or whatever limitations your embosser has. Do not use this approach if your turnaround schedule is too tight. Allow enough time to make sure all steps take place with breathing room if a problem occurs. The results of pulse embossing have been stunning and it is worth pursuing.

Flat integrals (a type of holographic stereogram) which we will discuss next, can also be embossed. This also is a relatively new option for embossed holography and opens up new territory for creative advertising.

Holographic Stereograms

A stereogram is defined in the dictionary as "a diagram or picture representing objects with an impression of solidity or relief". Consequently, holographic stereograms are holograms of diagrams or pictures which give the impresion of solidity or relief.

There are several techniques used to make holographic stereograms. Names are given to the various methods differentiating them and you will hear names like:

1. Integram
2. Cross hologram
3. Lesliegram
4. Multiplex hologram
5. Benton stereogram
6. Embossed stereogram
7. Alcove hologram
8. Holodisk®

These are all holographic stereograms. Items 1-5 can be considered the same type of stereogram. In order to give you a clear idea about what is going on, we will describe how a Cross holographic stereogram is made.

HOW TO MAKE A HOLOGRAPHIC STEREOGRAM

Generally, this is the way a 360-degree Cross holographic stereogram is made:

1) Make a small stage that rotates 360 degrees.

2) Put your subject on the stage.

3) Set up a regular movie camera on the stationary floor.

4) Film the subject as it turns the full 360 degrees. Slight motion is possible but the subject cannot make radical or jerky moves. This creates what is known as time smears in the hologram-places where the subject looks jagged. Slow, even movements when filming yield the best results.

5) Develop the movie film.

6) You now want to make a hologram of the image in each frame of your movie.

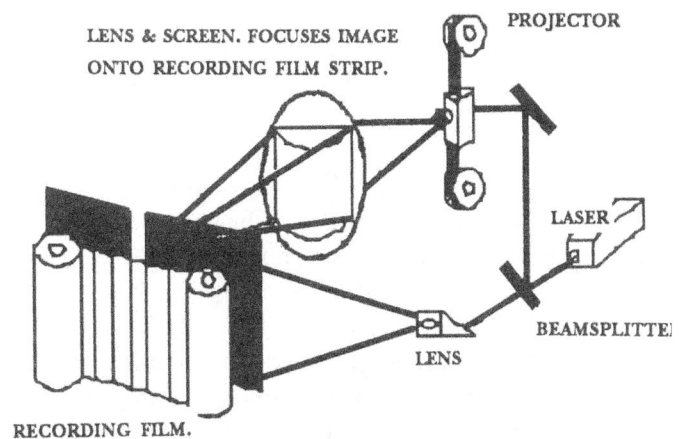

LENS & SCREEN. FOCUSES IMAGE ONTO RECORDING FILM STRIP.
PROJECTOR
LASER
BEAMSPLITTE
LENS
RECORDING FILM.

7) The holograms will be on a roll of holographic film. Each hologram, as you can see, will be as tall as

the width of the roll but of very narrow width.

8) Set up a mask in front of your roll of holographic film to get one narrow hologram (you will be shooting a white light rain bow transmission hologram).

9) Shoot the first exposure, advance the holographic film one frame, advance the movie film one frame, expose again and so forth for the entire movie film.

After development, you take the roll of holograms and wrap it around a strong cylinder of clear plastic that is mounted on a display which is able to rotate 360 degrees. Place a clear (unfrosted) light bulb with a single filament in the center just below the holographic film.

VIEWING THE HOLOGRAPHIC STEREOGRAM

Turn the light on and rotate the cylinder. Several things happen here:

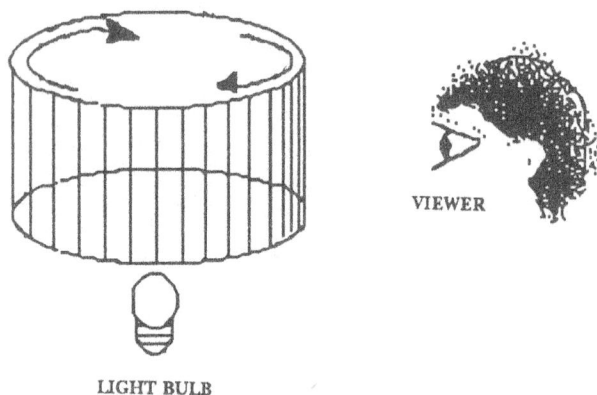

360 DEGREE HOLOGRAM
ROTATING CLOCKWISE

VIEWER

LIGHT BULB

1) The movie frames are moving past the eyes.

2) Each eye sees different images at the same time thus creating a stereo view.

3) Since this is a rainbow white light transmission hologram, the image is formed in the same position it was when originally shot (the image usually forms in the center of the cylinder).

What we have is a moving holographic stereogram and the viewer sees a three-dimensional "movie" of the image moving around in the cen ter of the cylinder.

OTHER HOLOGRAPHIC STEREOGRAMS
Less than 360-degree curoed Cross holographic stereogram

The holographic stereograms like the example above can be made in a curved format less than 360 degrees. On the market today are 90-, 120-, 180-, and 360-degree curved holographic stereograms. Those smaller than 360 degrees generally stay fixed and the viewer walks past them to see the motion. The smaller curved stereograms are inexpensive compared to the 36O-degree stereograms.

Flat holographic stereogram (Benton stereogram)

Instead of a curved stereogram as described above, you can make a flat holographic stereogram. The procedure for shooting is a little different.

The subject is not on a moving stage this time but on the ground at a distance from the camera. Now we create a straight railroad track on which to put the movie camera. Without aiming the camera directly at the subject, but facing the camera in the subject's direction, the camera moves along the track and takes photos at equal distances. We then develop the film and holograph it much the way we did in the Cross hologram described above.

The result is a flat sheet which, when held up to a light and tilted from side to side, displays an image in motion. With some changes in the holographic process, you could also make this a reflection hologram in which case the viewer simply stands under a light, holding the flat sheet and tilts it from side to side. The image can display above the surface of the hologram and move about as it is rocked from side to side.

Embossed holographic stereogram

Earlier it was men tioned that rain bow transmission holograms are used as the first step in the embossing process. Since a holographic stereogram is usually a transmission hologram, why not take a flat holographic stereogram and emboss it?

It turns out that this can be done and it is a popular application. Other techniques along this line include shooting live subjects with a pulse laser. Very recent developments allow holographers to reduce the size of the image in pulse shots. In theory you should be able to make a series of live shots, using a pulse laser, and reduce them to fit an embossed holographic stereogram. Perhaps someday there might be curved embossed holographic stereograms as labels for canned products.

Alcove holographic stereograms

What we have done above is to take some flat art., such as movie film, in which the subject moves from frame to frame and make holograms of it. As you know, computers can now make original flat art. Computer-generated images are obviously easier to work with, particularly when making corrections. Why not generate all the art you need to make a holographic stereogram using the computer?

This is the concept behind the alcove holographic stereogram. The group led by Steve Benton at the Massachusetts Institute of Technology's Holography Division of the Media Lab is working hard on this concept and they have already produced some remarkable results. The computer, in this case, creates the flat art. The flat art is then filmed from the video display terminal. The film is then used to make the holographic stereogram.

What also sets this hologram apart from the others is the way it is viewed. The image is seen within a clear, concave half-cylinder, an alcove, with close to 180 degrees of horizontal parallax. Along with the wide parallax, the image in the hologram can have depth going back through the hologram to infinity. Image distortions are corrected before transfer within the computer.

The alcove hologram is probably the closest holography has come to totally synthetic, three-dimensional subjects to date.

Holodisk®

This is a fascinating application of the reflection holographic stereogram. It is manufactured by Advanced Dimensional Displays. The process takes 360- degrees worth of photographic images and holographically integrates them onto a flat Holodisk®. It is lit from above and the subject can appear below, straddling, on, or above the disk surface within the limitations of the reflection hologram.

A slowly rotating turntable can bring 360-degrees of views to a stationary viewer. The obvious application would be to put the Holodisk® on phonograph records. As you listen to your favorite song an image

could dance before your eyes. Unfortunately the revolutions per minute are too high for the Holodisk® to be used on records, but research continues on this.

HOLOGRAPHIC STEREOGRAM SIZES

The sizes vary with each holographic stereogram. The standard curved Cross holographic stereogram is about ten inches high and eighteen inches in diameter with a curved face in 120-, 180- and full 360-degree formats. The widest roll film you can purchase is a standard 42 inches wide. Attaching one end of a roll to another, these holograms can theoretically be made in indefinite lengths. Using motion picture footage, we could conceivably run hundreds of feet of continuous holographic film. Current technical restrictions, however, need to be overcome before this can be a reality.

HOLOGRAPHIC OPTICAL ELEMENTS (HOE)

As we have seen, a transmission hologram behaves in some ways like a lens because it lets light pass through it and focuses that light to form an image. It is possible to make transmission holograms that act

HOLOSCAN

The World's most advanced mastering studio and laboratory.
Leading mass producer of world's largest holograms.
Network of illuminated holographic advertising sites throughout Europe.
Full design consultancy.

Head Office & Laboratory: East Tytherley Road, Lockerley, Hampshire SO51 0JT
Telephone: 0794 41229   Fax: 0794 41264
London Office: Telephone: 01-373 1878

as very good lenses. In other words, instead of making an image, the hologram is designed simply to spread or focus light. In the same way, reflection holograms can be made into exceptionally good mirrors. This is what the field of HOE is all about. We will not cover it in depth in this issue, but diffraction gratings are sometimes used by artists so we will say something about that now.

Diffraction Gratings

A diffraction grating is a device which bends light. Diffraction gratings can be made by etching, deposition, acoustically, or holographically.

Holographically, diffraction gratings are made by the interference of two (or more) beams of pure, undiffused laser light. Diffraction gratings are among the most simple holograms to construct and they may be designed to produce unusual bursts of very pure color or simply diffract a specific color. Diffraction gratings can also be made to reflect a very high percentage of the specific color you desire.

Diffraction gratings are used artistically. To create variations in the diffraction patterns, holographers pass the laser beam through different kinds of lens

pieces of plastic, etcetera. Reflecting the beam off surfaces like crumpled aluminum foil or even ground glass can yield surprising optical effects.

CHAPTER 2

ARTWORK

This chapter is going to discuss what kinds of artwork can be used for the subject you are going to holograph. The subject is going to be one of three things:

1) 2-D "flat art"-subjects like drawings, photographs, movie stills or computer animation.

2) 3-D subjects-objects you see in everyday life

3) A combination of 2-D and 3-D subjects

There are, of course, a lot of variations on each of the above. We will look at each of the possibilities.

2-D and 3-D Art

2-D FLAT OVERLAYS

When preparing 2-D graphics (anything which is flat art) for holographing there are several considerations. Any image can be holographed, providing it can be made stable for the exposure, and has the correct contrast under laser light. For this reason, black and white art is the most reliable, as colors read differently for exposure under the monochromatic laser light. If a red laser (Helium Neon gas laser) is used to record a color work, anything close to the red color of the laser will completely disappear in the recording. This principle can be demonstrated by looking through red acetate at color work.

When holographing flat art from a surface other than a transparency, the hologram records the artwork's edges which mayor may not be desirable. A three-inch square logo, black on white paper, eight inches behind the hologram recording plate records the paper's perimeter and the white background. If the visible background is not desired, it is best to have the flat image transferred to the correct size of an acetate. When doing this, be sure to transfer the image to acetate as opposed to a polyester or other clear, plastic material. Acetate tends not to depolarize the laser light passing through it. Polyester and other plastic materials often do.

Depolarization yields a hologram that is bright in some areas and dim or non-existant in other areas. As you look at a randomly depolarized hologram, it looks like there are dark clouds with light peeking through-making the image hard to see from different points of view. A clear acetate of flat images sandwiched in glass (or the more expensive-but more effective-black and clear graphics silkscreened directly onto glass) allows placement of images at any reasonable depth from the holographic recording plane.

There are several ways to holograph flat art. The most common is to back-light the acetate graphic using laser light through frosted glass. The holographic plate records every clear space in the acetate graphic as a glowing light, and the blackened parts as absence of glowing light-or black.

Letters, symbols or images can also be placed in front of a solid, white, smooth background or a background of sparkling glass dust. Lighting the acetate from the front, the black parts seem to hover above the lit background and they cast shadows onto the background. Shadowgrams adapt this technique. An object or flat art is back-lit with a screen of frosted glass in front. The result shows the object's silhouette each other. Transmission holograms display abundant subject information by simply layering two or three rain bow holograms in a sandwich for display.

It should be pointed out that, as in any process, the artwork is only as good as the artist. If you love what you have seen in embossed holography it is because some highly skilled artist made the artwork well, designing them with holography in mind. If you have hated what you have seen in embossed holography, first you probably have not looked at too many holograms and secondly, you might only have seen the results of talentless designers who underestimate the consuming public.

2-D MULTI-CHANNEL ANIMATION

Multi-channel animation can catch someone's attention and, on closer inspection can present several groups of visual information. These holograms have also affectionately been called "hologram winkies". They are not to be confused with the 2-D images that have numerous fine stripes coating the image- found in candy boxes as prizes for children.

Those are lenticular lens-coated images done with an entirely different line-screen process, and they are not holographic. We described multi-channels in the introduction.

3-D HOLOGRAM SUBJECTS

Holograms display true 3-D images of subjects and models, but if a CW laser is being used, the subjects must be extremely stable, because of the long exposure time.

Embossed holograms recently have been able to accommodate pulse laser holograms to their particular recording requirements. This is a breakthrough, bringing holography another step closer to the flexibilities of 2-D imaging media like photographyexcept in full 3-D! A pulse laser was used to shoot the masters for Geo magazine's cover of a live person, and more recently, for National Geographic's shattered globe cover.

Usually, models for transmission or reflection holo-

on a brightly lit screen.

The advantage of 2-D art is obvious. There is no depth to the object and therefore the hologram plays back under any lighting condition extremely well. Also, 2-D graphics usually affords the holographer better use of bright light and thus shorter exposures which give less chance of movement during the exposure. Very often, diffraction gratings are used as the 2-D graphics. They are very easy subjects to record and are very bright and sharp. Flat artwork is also less expensive to create and can use all existing forms of 2-D art from hand drawings and photographs to computer-generated graphics.

2-D SPATIAL OVERLAYS

Flat artwork can be exposed to display several layers of depth. The idea is that you have flat art placed at varying distances from the holographic plate. The effect is like a stage that has several backdrops at different distances from the audience. Space has to be left between the art to see all the layers. Even superimposing a small word on a background livens up a piece. The same word can also be made to "echo" off into space for a dynamic effect. Placing the graphics at different angles makes them appear to be intersecting

grams are painted a light neutral color to be as reflective as possible. A brass object is good for its solidity but would be a bad choice if highly polished to a bright shine. By coating the brass with a kind of vaseline substance, sometimes the shininess can be hampered, allowing the object to reflect diffuse light.

When recording a subject with a pulse laser, the same rules for surface color and quality apply as in CW recording. A person whose face is painted white is an easier subject to record, so some holographers have their models heavily made up to look like clowns, mimes and similar character types. Not everyone wants to or should do this. A bit of powder on shiny skin, some extra blue shading to define certain areas and the avoidance of red lips (they look more white under the red pulse laser light and tend to not be as clearly visible as lips) are some basic make-up techniques.

Recording different environments with a pulse laser is possible but if outdoors it must be shot at night and probably in some kind of enclosure. Even the earliest pulse hologram environment recordings are inspiring. The extra work and cost are well worth it. Any environment of the right colored and textured surfaces, up to 20 cubic feet, can be recorded accurately in pulse laser holography. Outdoor hologram subjects can also be shot on movie film and converted into a holographic stereogram.

2-D/3-D HOLOGRAM
You can combine both flat art and 3-D object in a hologram. The possibilities are unlimited . 2-D / 3-D multichannel holograms can display "before and after" or "before, during and after" images in two and three dimensions. A 2-D/ 3-D image combination for each hologram channel of viewing might display an old logo and product that shifts to the n ew logo and product.

Objects can also be holographed with different exposures to make them intersect or share the same space. They can even be holographed so that they will have different colors or will mix colors when they intersect. This is where the holographer can sculpt with light and produce stunning effects. The possibilities are limited only by the ad designer's imagination.

Reducing the Image

Recently, research in image reduction has progressed significantly. Up until two years ago, with the exception of holographic stereograms, holography gave a one-to-one size ratio of the object it records to the final hologram. In other words, holographing a four-inch flashlight, the hologram appears to be exactly four-inches tall. Holography does not use a lens like photography. When lenses were used to shrink or enlarge an object or graphic, optical distortion occurred which made this process unpopular. When reducing objects holographically, there was a proportional reduction in the depth of field as well. This also was undesirable to hologram producers.

In the last two years, there has been significant progress in image reduction. The advancement of pulse portrait use has created a demand for reducing large pulse hologram images. One goal of the research is .to create a size and scale of subjects that is psychologically closer to photography's scale in publications and on television.

A person's portrait in full 3-D, life size, is scary to a surprising number of people just as the first photographs and motion pictures were. Reductions of pulse portrait images still look very real but are too small to actually be that person.

One reason research on image reduction is progressing is because the shrinking of depth of field and of the image makes it possible to easily transfer a pulse hologram image to the standard sixinch square format of embossed holograms. The embossed hologram is the major commercial massmanufacturing format for holograms. One could take holographic portraits of famous people, reduce the portraits and emboss them in the millions.

CW holograms can be reduced, but usually are not. Objects must be still, so usually models are made. If the model has to be built then it is scaled down or up at the same time it is made.

Recording Material

There are two types of recording material, film (or polyester) base and glass. These bases are called substrates. Film and polyester are more flexible in size and shape than glass. Standard sizes are available in flat sheets and rolls. The flat sheets are often easier for hologram producers to work with because they are less likely to curl. Film and polyester bases are also less expensive and more responsive to low levels of laser power, allowing for shorter exposures. The largest size of silver halide film is the forty-two inch wide rolls that are ten to thirty meters long.

Glass plate substrates are also available for holographic recording materials. Standard sizes are as small as three square inches and as large as 30 cm x 40 cm in the U.S. and just slightly larger in Europe. Custom sizes can be made but they are expensive. Standard sizes on plates are already on the average of five times more expensive than film or polyester substrates. Hologram producers use the glass base material for H-I master recording in holography because of the stability and the even surface. Glass may also out-last the others for archival purposes; however, it is less reliable in consistency of emulsion hardness. Figuring new exposure times and fussing with sensitivity treatments is time-consuming and expensive. Some holograms of the square meter size exist on glass, but these are no longer a stock product, unfortunately.

CHAPTER 3

CONTRACTS

We assume that you have decided you want to produce a hologram. The next thing to do is sit down and write out a first draft contract or purchase order. It is amazing how many things are cleared up when you spell out in detail all the items required for a project

One thing you should recognize is that holography is a new industry. The first mass-manufactured, embossed hologram published in a commercial trade book was in late 1982. This tells you two things.

1) You can't look for old, established manufacturers because there aren't any; although some manufacturers are better than others.

2) You must have a very specific contract and write out every possible detail of your project. No sensible business person would order any product without written and signed communication. In an industry like holography, where most of the technology is new, it is imperative to do this.

Every company has its own legal contracts and purchase orders so the best we can do is make a list of items that you should be aware of in making your contract. Here are the most important items:

Detail the Project

Before doing anything, make sure you know what you want. Have a clear idea of what you want in the hologram and when and how it is going to be shown. Is it embossed and applied to a magazine? Specifically, what is the subject of the hologram?

The point is obvious: if you are fuzzy about what the

project is, the manufacturer will be equally unclear. If everybody is unclear, then the project suffers. You should freely call holographers and manufacturers for answers to your questions and while we encourage you to do this, please be aware that many are very busy. Consequently, they are prone to charge for their time as consultants. These consulting fees are probably worthwhile if they save you a lot of research time.

Compare Prices and Services

It will take a few phone calls to settle the project specifications in your mind and on paper. As soon as you have a clear set of specifications you can call for price quotes using the same specifications for each vendor. There are some salespeople in this industry that will try to make you think their company is the only one with which you should do business. Call their competitors and you will get a different story. Ask for samples and references and call these people up. Ultimately it is the references, quality and price of the work you have seen that will assure that the hologram you want will be made correctly, delivered on time and at a fair price.

Creation of Subjects

There are techniques available that allow you to reduce the image size, but these techniques are new and expensive. The more common route is that the subject is the same size as your final hologram (either 2-D artwork, or a 3-D object). In either case, before spending any money, consult the holographer you have hired to do the shot and the manufacturer to make sure you are creating the correct model. It usually does make a difference what color and materials your subject is made of. Write all this in the contract so there are no "that is not what I said" excuses later. It is also advisable to take time to find a very good artist For the kind of money you will be spending, the extra cost for a good artist is negligible. The hologram itself will only look as good as the object it is taken from. Also, the cost of a bad artist is more than just money since your name is attached to the product

Proofs

You want a lot of these. It is your only way of being assured that each major step has been completed satisfactorily. Establish in the contract several points at which you will see proofs and approve the project's continuance. The fewest number of proofs would require:

a) inspecting the model when completed. Remember that what-you-see-is-what-you-get and the

hologram is not going to make a bad model look good.

b) proofing the master before mass-manufacturing. Remember that you lose a little quality going from master to mass-manufacturing so you should be fairly happy with what you see.

What happens if (a) or (b) is unacceptable? Write it out in detail.

Multiple vendors

If two or more vendors are involved, such as the model maker, holographer, embosser, and printer, then specifications (in writing) should be provided by each vendor for what each will accept. Everyone in the project should see, and agree in writing on each others' specifications. This is also a good place for proofing. When the project is passed from one vendor to the next, there should be something in writing that says the project was received by vendor X and it meets all specifications laid out in the contract.

Payments

This is a standard part of any contract. It should be spelled out clearly so you don't wind up with an unexpected demand for cash that stalls the project.

Copyright

Holograms are regarded by the law as works of art. Consequently, the images can be copyrighted. To copyright a hologram image in the USA, you call the copyright office at (1) (202) 4790700. Ask to be sent a "VA" form for copyright. You then return the completed form with the $10 fee, and either a sample of the work of art (the hologram) or a picture of it. You receive in return an official certificate of copyright. When we called the copyright office, they pointed out that the way in which the hologram is mounted or displayed is sometimes considered part of the work of art; and therefore, sending a sample hologram as well as a picture is recommended.

Remember to include in your contract with manufacturers a detailed statement saying that you hold the copyright and copies are not to be made without your written agreement. If everyone involved in the project signs a contract clearly stating that you hold the copyright, then any extra copies finding their way to the public will give you cause to complain.

Royalties

If you are going to give royalties to an artist, it is best to have a separate royalty contract made between you and the artist where the artist becomes the copyright holder (you will copyright the work in the artist's name) but the artist relinquishes to you the right to manufacture and sell the artwork. You should write out in the royalty contract how long it is in effect and what areas of the world are covered.

If you are simply paying an artist a flat fee to do a model or hologram for you, be careful. It is quite possible to give an artist free reign to create a model or hologram according to your general directions and later find out that the artist assumes he owns the copyright and expects some royalties because it was his artistic creation.

If you are using more than one artist, it is also a good idea to put in writing what artwork in what form is to be made and who will ultimately be the art director for the project.

Deadlines

Without deadlines, finishing your project is improbable. Establish in the contract deadlines for each major step. What happens if the deadlines are not met? Write it down in detail.

Termination of contract

What happens if you don't like the product you are getting and decide to terminate the contract at some stage? Perhaps you may even decide you want to use another vendor and pull your job from the person now doing it. Do you have provisions in the contract that state that all work to date is your possession and is to be turned over to you immediately if you decide to terminate the contract? Are any of the vendors going to be able to charge you for things not yet donemaybe they ordered special supplies for you? Write it all down in detail.

Masters

There is generally more than one master made because you do not want to have to go back and do the original shot over if something happens to your first master. How many masters will be made? How much will they cost? Write it down.

Hourly time

Sometimes you will come to services, such as creating a model, where the vendor doesn't really know how long it will take and wants to charge you by the hour. You should then ask for a "not to exceed so much money" clause. Most good business people will be willing to give you a "not less than this much or more than this much" quotation in writing.

Reprints

After the project is over, suppose you decide to go back to press and run off a thousand more copies. How much will it cost? Does the manufacturer save the masters for you or is it that your responsibility?

Finishing work

It is easy to figure out all the manufacturing costs and forget the packaging costs. How is your hologram going to be displayed? Is the manufacturer going to do any of the packaging? If so, you need to write it all down in detail. Visualize how your final product looks and ask yourself who is making each of the final display parts.

Summary

It is impossible for us to think of all the things that should be written into in a contract but we hope the points above might help you a little. Most of the businesses in the holography industry are run by honest people and they want to do the best possible job for you. Problems usually arise from bad communication, not dishonesty. Remember that a short contract is easy to write but difficult to enforce.

Include your company!

Send information about your holography services or products to:

HMP Editor
Ross Books
P.O. Box 4340
Berkeley, Calif.
94704 USA

Telephone: (1)(415) 841 2474

FAX: (1)(415) 841 2695

CHAPTER 4

PRODUCERS

Producer Chart

In this section we give you a list of the businesses that manufacture or produce holograms. This covers a very wide range of activities: from the speciality limited edition holographers to the embossed manufacturers. If you are unclear about what a specific service is, consult the introduction.

The chart lists all the companies and checks which producer categories they fit into. This allows you to scan through all the companies for the area of your interest. Branch offices have separate listings, so there may appear to be duplicates, but their addresses are in fact different. Companies are listed alphabetically within their country.

We sent questionnaires to all producers. This helped us update the list quite a bit but we did not receive questionnaires back from everyone. We will do this again next year and again eliminate those not responding in order to give you a very clean and updated list.

If you would like to have us list new categories or if you have any other suggestions please write us. We do read the mail and will try to improve the book yearly.

Producer Names and Addresses

This section lists the full company name and address, telephone, fax and telex. All categories which the company has checked off-in Producer, Buying & Selling, Holography in Education, and Equipment & Supplies areas-are listed to give you a better idea of the scope of their activities. As in the chart section, companies are listed alphabetically within their country.

While we have attempted to print the correct country and city codes for international telephone and fax contact, some codes may be missing because we were unable to verity them with the international operator in time for publication. If your country or city code is incorrect or missing, please notify us.

HOLOGRAPHY PRODUCERS

AUSTRALIA — CANADA

	A1 COMMERCIAL H-1	A2 FINE ART ORIGINALS	A3 FINE ART LTD EDIT'S	A4 CUSTOM PRESENTATION SUPPT	A5 HOLOGRAPHIC STEREOGRAMS	A6 EMBOSSED, HOT-STAMPED	A7 SILVER HALIDE TRANSMISSION	A8 SILVER HALIDE	A9 CUSTOM LIGHTING SUPPORT	A10 OVERSIZE LARGE FORMAT	A11 PULSE HOLOGRAMS	A12 DICHROMATE
AUSTRALIA												
orge Gittoes	✓											
ser Light Expressions Pty. LTD.	✓	✓	✓	✓	✓	✓	✓	✓	✓	✓		
ser Electronics Pty. Ltd.	✓						✓	✓				
zart Pty. Ltd.	✓											
linda Menning	✓											
n Pye	✓											
nis Quinlan	✓											
BELGIUM												
re Boon	✓											
ologram Europe sprl.	✓											
ger Graphic Technology	✓											
BRAZIL												
olografica	✓											
CANADA												
udette Abrams	✓	✓										
tec Engineering Limited	✓										✓	
sociates of Science and Tech	✓							✓				
ryn Cadell	✓											
rie-Andree Cossette	✓	✓										
lissa Crenshaw	✓											
ep Space Holographics	✓											
lney Dinsmore	✓											
borah Duston	✓											
orges M. Dyens	✓		✓									
ques Frigon	✓											
nge Research Holographics	✓	✓				✓	✓	✓			✓	✓
neral Holographics, Inc.	✓											
ma Heaton	✓											
locor IBF. Printing Inc.	✓				✓							
locrafts	✓											
lo-Dimensions Inc.	✓											✓
lo Laser Tech	✓											

HOLOGRAPHY PRODUCERS

	A1 COMMERCIAL H-1	A2 FINE ART ORIGINALS	A3 FINE ART LTD EDIT'S	A4 CUSTOM PRESENTATION SUPPT	A5 HOLOGRAPHIC STEREOGRAMS	A6 EMBOSSED, HOT-STAMPED	A7 SILVER HALIDE TRANSMISSION	A8 SILVER HALIDE	A9 CUSTOM LIGHTING SUPPORT	A10 OVERSIZE, LARGE FORMAT	A11 PULSE HOLOGRAMS	A12 DICHROMATE
ANADA (cont'd) — HUNGARY												
olomorph Visuals, Inc	✓											
lospectra Inc. (Ste-Foy)	✓											
lospectra Inc. (Ste-Therese)	✓	✓	✓									
OU Inc.	✓	✓	✓	✓		✓	✓	✓	✓			
ik Lafreniere	✓						✓	✓				
ser Holographics, Inc.	✓					✓						
ser Innovations Inc.	✓											
val University (Quebec City)	✓	✓	✓				✓	✓				
s Productions Hololab!	✓						✓			✓		
ght Construction, Inc.	✓	✓	✓									
n-Pierre Marchand		✓										
3D Modulations		✓										
bert Myre	✓											
nne Roy	✓											
rnd Simson	✓											
ilone Holographie Corp.	✓	✓	✓	✓		✓	✓	✓	✓	✓	✓	
DENMARK												
thioff Johansen		✓	✓									
FRANCE												
F PRODUCTIONS	✓											
n Gilles	✓											
olographic Creations	✓											
ologram. Industries	✓											
lo - Laser	✓	✓	✓	✓		✓	✓	✓		✓		
lo-Visual Concept 3D	✓	✓	✓	✓				✓		✓		
HOL	✓											
tical Laboratory	✓											
nis Palamas	✓											
ques Senechal	✓											
ouis Tribillon	✓											
AL.	✓	✓	✓	✓		✓	✓	✓				✓
HUNGARY												
tplay Holographic Studio	✓	✓	✓									

HOLOGRAPHY PRODUCERS

ISRAEL — SWEDEN	A1 COMMERCIAL H-1	A2 FINE ART ORIGINALS	A3 FINE ART LTD EDIT'S	A4 CUSTOM PRESENTA- TION SUPPT	A5 HOLOGRAPHIC STEREOGRAMS	A6 EMBOSSED, HOT- STAMPED	A7 SILVER HALIDE TRANSMISSION	A8 SILVER HALIDE	A9 CUSTOM LIGHTING SUPPORT	A10 OVERSIZE LARGE FORMAT	A11 PULSE HOLOGRAMS	A12 DICHROM- ATE
olofar Lab Israel	✓											
ITALY												
torio Alliata	✓											
nbolegge & Bimbogioca SRL	✓											
olofar Lab (SRL)	✓											
JAPAN												
toh & Company	✓											
lo ARP	✓	✓	✓	✓		✓						
tsuko Ishii	✓	✓										
miko Shiozaki		✓										
unsuke Mitamura	✓											
LUXEMBOURG												
vid Dewar	✓											
MEXICO												
ologramas De Mexico	✓	✓		✓	✓	✓		✓	✓			✓
NETHERLANDS												
exander Coblijn	✓											
itch Holographic Laboratory	✓		✓	✓	✓	✓	✓	✓	✓	✓		✓
undation Ideecentrum	✓											
e Sera Sera	✓											✓
alter Spierings	✓			✓	✓	✓			✓			✓
hite Tiger Holograms	✓											
SOUTH AFRICA												
olotec CC	✓	✓	✓	✓			✓	✓				
SPAIN												
toni Pinol Gullamon	✓											
olotek	✓											
idimensionale Hologramas	✓		✓									
SWEDEN												
rgny E. Carlsson	✓											
A. Jonsson	✓											
sergruppen Holovision AB	✓									✓	✓	
arwell AB	✓	✓					✓	✓		✓		

HOLOGRAPHY PRODUCERS

	A1 COMMERCIAL H1	A2 FINE ART ORIGINALS	A3 FINE ART LTD EDIT'S	A4 CUSTOM PRESENTATION SUPPT	A5 HOLOGRAPHIC STEREOGRAMS	A6 EMBOSSED, HOT-STAMPED	A7 SILVER HALIDE TRANSMISSION	A8 SILVER HALIDE	A9 CUSTOM LIGHTING SUPPORT	A10 OVERSIZE, LARGE FORMAT	A11 PULSE HOLOGRAMS	A12 DICHROMATE
...ITZERLAND — UNITED ...NGDOM												
...cal Barre	✓							✓				
...erart Ford AC.	✓											
...l Fredrick Reutersward	✓											
...TAIWAN												
...ng Ling Industrial Research	✓											
...UNITED KINGDOM												
...vanced Holographics, Ltd.	✓	✓	✓	✓		✓	✓	✓	✓	✓		
...l, Prismatic, Inc.							✓	✓				
...tt Andrews	✓											
...plied Holographics		✓	✓				✓	✓				
...ot Laser Picture Studio		✓	✓				✓	✓				
...r & Stroud, Ltd.	✓	✓					✓	✓				
...rgaret Benyon	✓	✓										
...rick Boyd	✓	✓			✓						✓	
...odel Holograms	✓	✓			✓						✓	
...er Buxton	✓											
...gela Coombes	✓											
...an Ann Cowles	✓		✓									
...rkroom Eight Ltd.	✓											
...rersified Optical Ltd.	✓											
...sound Ltd.	✓											
...k Hardy	✓											
...n Harris	✓											
...ne Hicknott	✓					✓						
...olofax Limited	✓											
...logram One	✓							✓		✓		✓
...olographics (UK) Ltd.	✓	✓		✓			✓	✓	✓	✓	✓	
...oloprint Ltd	✓			✓			✓	✓	✓	✓	✓	
...oloscan Ltd	✓						✓	✓				
...olotec PLC	✓							✓				
...olovision Ltd..	✓							✓				
...thony Hopkins	✓										✓	
...: Brown Holographics	✓											

HOLOGRAPHY PRODUCERS

UNITED KINGDOM (cont'd)	A1 COMMERCIAL H-1	A2 FINE ART ORIGINALS	A3 FINE ART LTD EDIT'S	A4 CUSTOM PRESENTATION SUPP'T	A5 HOLOGRAPHIC STEREOGRAMS	A6 EMBOSSED, HOT-STAMPED	A7 SILVER HALIDE TRANSMISSION	A8 SILVER HALIDE	A9 CUSTOM LIGHTING SUPPORT	A10 OVERSIZE LARGE FORMAT	A11 PULSE HOLOGRAMS	A12 DICHROMATE
ris Lambert	✓											
ser Lines Ltd.	✓											
za Holograms	✓	✓			✓		✓	✓			✓	
ris Levine	✓											
ght Fantastic Ltd.	✓	✓	✓	✓	✓	✓	✓	✓	✓	✓	✓	✓
ght Fantastic PLC	✓	✓	✓	✓	✓	✓	✓	✓	✓	✓	✓	✓
ght Impressions Europe, Ltd	✓					✓						
ghtworks	✓											
iane Lijn												
drew Logan												
chael Medora												
rage Holograms Ltd.	✓	✓	✓	✓	✓		✓	✓			✓	
hit Mistry												
nday Spatial Imaging	✓	✓	✓	✓		✓	✓	✓				
wbold Wells Company	✓	✓	✓	✓			✓	✓				
ul Newman	✓											
-Graphics (Holography) Ltd.	✓	✓	✓	✓		✓	✓	✓	✓	✓	✓	✓
tical Engineering Group	✓	✓	✓	✓			✓	✓	✓		✓	
tical Surfaces Ltd.	✓	✓	✓									
tical Works Ltd.	✓											
iel Scientific Ltd.	✓											
ford Holographics	✓											
ford Scientific	✓	✓	✓									
roline Palmer												
drew Pepper	✓		✓					✓				
rception Holography	✓	✓					✓	✓	✓		✓	
kington P.E. Ltd.	✓											
vid Pizzanelli	✓	✓	✓	✓		✓	✓	✓	✓			
yen Holo Ltd.	✓	✓	✓	✓	✓		✓	✓	✓		✓	
chmond Holographic Studios	✓		✓							✓		
lfin Ltd.	✓											
sewell Ltd.	✓											
ope Optics Ltd.	✓					✓					✓	

HOLOGRAPHY PRODUCERS

Producer	A1 COMMERCIAL H-1	A2 FINE ART ORIGINALS	A3 FINE ART LTD EDIT'S	A4 CUSTOM PRESENTA-TION SUPPT'	A5 HOLOGRAPHIC STEREOGRAMS	A6 EMBOSSED, HOT-STAMPED	A7 SILVER HALIDE TRANSMISSION	A8 SILVER HALIDE	A9 CUSTOM LIGHTING SUPPORT	A10 OVERSIZE, LARGE FORMAT	A11 PULSE HOLOGRAMS	A12 DICHROM-ATE
UNITED KINGDOM (cont'd) –												
...ques Senechal	✓											
...ck Silberman	✓											
...ECAC Limited	✓											
...ectrolab Limited	✓											
...chNorth	✓											
...ird Dimension Ltd.	✓											
...uchwood Holographic Studio	✓									✓		
...aham Tunnadine												
...ichael Waller-Bridge	✓											
...enyon & Gamble		✓										
...ptan Lamps Ltd	✓											
UNITED STATES OF AMERICA												
...bott Laboratories	✓	✓		✓	✓	✓	✓	✓	✓			✓
...0 2000												
...lvanced Dimensional Displays												
...lvanced Environmental Research												
...H. Prismatic Inc.						✓	✓	✓				
...tes Lightworks		✓	✓				✓	✓				
...lexander"		✓	✓				✓	✓				
...nerican Bank Note Holographic						✓				✓	✓	
...nherst Media	✓											
...ait Studio		✓	✓	✓								
...pplied Holographics, Corp.	✓					✓	✓	✓			✓	
...mstrong World Industries	✓										✓	
...tigliography	✓	✓					✓	✓				
...tKitek												
...le Aust.												
...iss, Barefoot and Associates.		✓										
...eph Belk.		✓										
...die Berkhout			✓									
...berta Booth Studio	✓											
...y Bradshaw	✓											
...ookhaven National Laboratory.	✓											

HOLOGRAPHY PRODUCERS

A. (continued)	A1 COMMERCIAL H-1	A2 FINE ART ORIGINALS	A3 FINE ART LTD EDIT'S	A4 CUSTOM PRESENTATION SUPPT	A5 HOLOGRAPHIC STEREOGRAMS	A6 EMBOSSED, HOT-STAMPED	A7 SILVER HALIDE TRANSMISSION	A8 SILVER HALIDE	A9 CUSTOM LIGHTING SUPPORT	A10 OVERSIZE LARGE FORMAT	A11 PULSE HOLOGRAMS	A12 DICHROMATE
k Bunts		✓										
ph Burns, Jr.	✓	✓										
ard Bush	✓	✓										
neth G. Byrne		✓										
bridge Stereographics Group					✓							
d Carlton		✓	✓	✓	✓		✓	✓	✓	✓	✓	
lin-Silver Holography (Boston)		✓	✓	✓	✓		✓	✓	✓	✓	✓	
lin-Silver Holog. (Brookline)		✓	✓	✓			✓	✓	✓	✓		
rry Optical Company	✓	✓										
urn Corporation						✓						
e Conner		✓										
y Connors	✓	✓										
ion Corp.	✓											
hael E. Crawford		✓										
wn Roll Leaf, Inc.						✓						
hael Croydon		✓										
mas J. Cvetkovich		✓										
k Davis		✓										
zle Enterprises, Inc.		✓				✓						
k DeFreitas		✓										
enne B. Derichs		✓										
cent DiBiase												
ard Dietrich												
raction Company, Inc.						✓						
nensional Imaging Technology			✓	✓	✓	✓	✓					
tension Research	✓	✓										
nglass Associates Studios	✓	✓										✓
n Drevenak												
ip Dubov		✓										
Company	✓	✓		✓		✓						
tman Kodak Company	✓											
ith Eisen	✓	✓										
M												

HOLOGRAPHY PRODUCERS

.S.A. (continued)	A1 COMMERCIAL H-1	A2 FINE ART ORIGINALS	A3 FINE ART LTD EDIT'S	A4 CUSTOM PRESENTA-TION SUPP	A5 HOLOGRAPHIC STEREOGRAMS	A6 EMBOSSED, HOT-STAMPED	A7 SILVER HALIDE TRANSMISSION	A8 SILVER HALIDE	A9 CUSTOM LIGHTING SUPPORT	A10 OVERSIZE, LARGE FORMAT	A11 PULSE HOLOGRAMS	A12 DICHROM-ATE
atiron Studio												
sannah D. Foltz		✓										
thur David Fornari		✓	✓									
t Freund							✓	✓				
olographic Studios		✓	✓	✓		✓	✓	✓	✓	✓		
2M		✓		✓	✓	✓	✓	✓	✓	✓	✓	✓
eidi S. von der Gathen		✓				✓						
obal Images	✓											
Going	✓											
uce Goldberg												
nneth M. Goss												
atthew E. Hansen												
TI - High Tech Imaging	✓											
oechst-Celanese Corp.												
olage	✓	✓	✓	✓	✓	✓	✓	✓				
olaxis Corporation	✓	✓	✓				✓	✓	✓	✓	✓	✓
olicon Corporation	✓	✓	✓				✓	✓	✓	✓	✓	
olicon		✓					✓	✓	✓	✓	✓	
olocraft International	✓											
olocrafts of Long Island	✓											
oloflex Company	✓		✓		✓	✓	✓	✓				
olography Institute	✓											
olografix (Piedmont)	✓											
olograma	✓											
olographic Applications	✓	✓	✓	✓	✓	✓	✓	✓	✓	✓	✓	✓
olographic Design, Inc.	✓	✓	✓	✓		✓						✓
olographic Design Systems Inc	✓			✓	✓	✓	✓	✓	✓	✓		
olographic Dimensions, Inc.	✓											
olographic Images, Inc.	✓											
olographics,Inc.	✓	✓										
olographics North, Inc.	✓											
olography News	✓	✓		✓								
olography Workshops	✓						✓	✓		✓	✓	✓

HOLOGRAPHY PRODUCERS

U.S.A. (continued)	A1 COMMERCIAL H-1	A2 FINE ART ORIGINALS	A3 FINE ART LTD EDIT'S	A4 CUSTOM PRESENTATION SUPPT	A5 HOLOGRAPHIC STEREOGRAMS	A6 EMBOSSED, HOT-STAMPED	A7 SILVER HALIDE TRANSMISSION	A8 SILVER HALIDE	A9 CUSTOM LIGHTING SUPPORT	A10 OVERSIZE LARGE FORMAT	A11 PULSE HOLOGRAMS	A12 DICHROMATE
o-Spectra				✓		✓	✓	✓	✓			✓
owaves Hawaii	✓		✓					✓				
rd Photo Corporation	✓											
ges Company	✓											
ging & Design		✓	✓	✓		✓	✓	✓	✓	✓	✓	
itute of Optical Research							✓					
dy James/Holography		✓										
n Kaufman Photography	✓	✓						✓		✓		
on Kurzen		✓										
art Inc.	✓											
art Ltd.	✓											
er Arts		✓	✓		✓		✓	✓	✓			
.S.E.R. Co.		✓	✓				✓	✓	✓	✓		
er Dreams		✓		✓								
er Fare Ltd.		✓										
er Ionics Inc.		✓										
ermedia	✓	✓										
e Lasersmiths	✓		✓	✓						✓		
er Affiliates		✓	✓				✓	✓	✓			
da Law Holographics		✓	✓				✓	✓	✓			
ristopher J. LeSar		✓	✓									
wen Lev		✓										
onix		✓	✓									
ht Harmonics Inc.	✓				✓		✓	✓	✓	✓	✓	
l Light	✓					✓						
ht Impressions Inc.	✓					✓						
al Lubetsky		✓										
cShane Holography		✓										
n Environment, Inc.		✓										
rald Marks Studio	✓	✓	✓		✓							
nes A. McClean		✓		✓								
n McNair		✓										
dia Interface, Ltd.	✓			✓					✓			

HOLOGRAPHY PRODUCERS

S.A. (continued)	A1 COMMERCIAL H-1	A2 FINE ART ORIGINALS	A3 FINE ART LTD EDIT'S	A4 CUSTOM PRESENTA-TION SUPPT.	A5 HOLOGRAPHIC STEREOGRAMS	A6 EMBOSSED, HOT-STAMPED	A7 SILVER HALIDE TRANSMISSION	A8 SILVER HALIDE	A9 CUSTOM LIGHTING SUPPORT	A10 OVERSIZE, LARGE FORMAT	A11 PULSE HOLOGRAMS	A12 DICHROMATE
chael G. Merrick		✓										
rk C. Merrill		✓										
ve Moore		✓										
ron Muskovitz		✓										
chael Naimark		✓										
eovision Productions	✓	✓	✓									
w York Holographic Lab	✓					✓	✓	✓				
rth American Holographics	✓						✓	✓				
hner Holographics												
cific Holographics, Inc.												
R. Padnos				✓			✓	✓	✓			
el Petersen		✓										
tonio Peticov	✓	✓										
int of View Dimensions	✓	✓	✓	✓	✓	✓	✓	✓		✓		✓
laroid Corporation	✓		✓	✓			✓	✓	✓	✓		✓
rtson Inc.	✓											
r. Richard Rallison	✓	✓										
connaissance	✓	✓										
gal Press Inc.	✓	✓										
an Rhody	✓	✓										
el Rootstein, Inc.		✓	✓	✓			✓	✓				
ichael P. Rosewell		✓										
. Schultz & Co.		✓										✓
bert Sherwood Holog. Design	✓	✓				✓						
therine J. Smith	✓	✓										
cha Sonis												
ace Age Designs	✓	✓			✓	✓						✓
edratek Corp	✓	✓										
ve Provence Holography												
n Stockler		✓				✓						
nchronicity Holograms								✓				
ird Dimension Arts, LTD												✓
nald K. Thornton		✓										

HOLOGRAPHY PRODUCERS

	A1 COMMERCIAL H-1	A2 FINE ART ORIGINALS	A3 FINE ART LT'D EDIT'S	A4 CUSTOM PRESENTA-TION SUPP'T	A5 HOLOGRAPHIC STEREOGRAMS	A6 EMBOSSED, HOT-STAMPED	A7 SILVER HALIDE TRANSMISSION	A8 SILVER HALIDE	A9 CUSTOM LIGHTING SUPPORT	A10 OVERSIZE, LARGE FORMAT	A11 PULSE HOLOGRAMS	A12 DICHROM-ATE
S.A. (continued) - [U.]EST GERMANY												
...nsfer Print Foils, Inc.						✓						
...o Guys Holography		✓										
...uglas E. Tyler		✓										
...l Vance		✓										
...c Van Hamersveld		✓										
...ris Vila	✓	✓	✓	✓			✓		✓	✓		
...ve Guides Inc.	✓	✓	✓	✓	✓		✓	✓	✓	✓		
...ly Weber		✓	✓							✓		
...fford B Weissman												
...ndrajean Wells		✓										
...Wesly							✓	✓				
...ite Light Works, Inc.	✓	✓	✓	✓	✓	✓	✓	✓			✓	✓
...holeHogGraphy		✓			✓		✓	✓				
...hole Message Studios												
...rol Yeager	✓											
...e Zeman	✓											
WEST GERMANY												
...tlev Abendroth	✓											
...beitskreis Holografie B.V.	✓											
...omika Technische Physik GmbH	✓											
...rmenza Dominguez	✓											
...ian Fischer	✓											
...rtin Hofmann	✓											
...lar Seele KG	✓											
...lografie - Hofmann Labor	✓	✓	✓	✓	✓	✓	✓	✓	✓	✓	✓	
...OLO GmbH	✓	✓	✓	✓	✓		✓	✓	✓	✓		✓
...lographie Labor	✓	✓	✓				✓	✓		✓		✓
...ter Jung	✓											
...dreas Kaufman	✓											
...bor fur Holografie	✓											
...serion Handels GmbH	✓											
...lo Laube	✓											
...rlin	✓											

HOLOGRAPHY PRODUCERS

	A1	A2	A3	A4	A5	A6	A7	A8	A9	A10	A11	A12
	COMMERCIAL H1	FINE ART ORIGINALS	FINE ART LTD EDIT'S	CUSTOM PRESENTA-TION SUPPT.	HOLOGRAPHIC STEREOGRAMS	EMBOSSED, HOT-STAMPED	SILVER HALIDE TRANSMISSION	SILVER HALIDE	CUSTOM LIGHTING SUPPORT	OVERSIZE, LARGE FORMAT	PULSE HOLOGRAMS	DICHROM-ATE
EST GERMANY (continued)												
...ndeler & Hoyer GmbH	✓											
...inbichler Labor	✓											
...rau Volker	✓											
LATE ENTRY: SWEDEN												
...lo Media AB/Hologram Museum	✓					✓					✓	✓

Holo ArP

THE ONLY TOTAL HOLOGRAPHIC SERVICES
IN JAPAN

"HOLOGRAPHIC"

FOR PRINTER
FOR ART PROJECT
FOR EXHIBITION
FOR ARTISTS
FOR BUYING + SELLING
FOR EDUCATION
FOR EQUIPMENT + SUPPLIES

Holo ArP

㈱カマクラ・ホログラフィー事業部
No.1007 2-4-5 Shiba-Daimon Minato-ku Tokyo, 105 Japan Tel. 03-574-8307 Fax. 03-574-8377 〒105 東京都港区芝大門2-4-5 芝ダイヤハイツ1007
KAMAKURA GALLERY GROUP

HOLOGRAM PRODUCERS

A1= Commercial H-1	**A5**=Holographic Stereograms	**A9**= Custom Lighting Support
A2=Fine Art Originals	**A6**=Embossed/Hotstamped	**A10**=Oversize/Large Format
A3=Fine Art Ltd Editions	**A7**=Silver Halide Transmission	**A11**=Pulse Holograms
A4=Custom Presentation Support	**A8**=Silver Halide Reflection	**A12**=Dichromate

AUSTRALIA

George Gittoes
54 Brighton Street
Bundeena
NSW
2230 Australia
Telephone (61) (2) 523197
Categories Al

Laser Light Expressions Pty. Ltd.
G 12 Denawen Ave. Castle Grove
P.O. Box 23 Chatswood 2057
N.S.W., Australia
Telephone (61) (612) 4072896
Contact: J.T ohin, Rosie
FAX: (61)(612)4177247
Categories: A1 A2 A3 A4 A5 A6 A7 A8 A9 A10 B2
B3 B5 B6 B7 B8 B9 C4

Laser Electronics Pty., Ltd.
P.O. Box 359
Southport
Queensland, 4215
Australia
Telephone (61)(75) 321 699
Categories: A1

Lazart Pty. Ltd.
22 Erina Valley Road
Erina, New South Wales
2250, Australia
Telephone (61) (43) 676245
Contact Roslyn Wlison
Categories: A7 A8 B1 B2 B3

Melinda Menning
171 Hopetown Avenue
Vaucluse 2030
Sydney N.S. W.
Australia
Telephone (61) (2) 3371916
Categories Al

TimPye
19 Wonderland Avenue
Tamarama
Sydney
2026 Australia
Telephone (61) (2) 306 611
Categories: A1

Denis Quinlan
48 Clifton Street, North Balwy
Victoria 33104
Australia
Telephone (61) (3) 8578655
Categories: A1

BELGIUM

PiereBoon
St Pietersnieustaraat 41
GentB9000
Belgium
Telephone (32) (91) 233821
Categories: Al A11

Hologram Europe sprl.
Avenue Voltaire 137
1030 Brussels
Belgium
Telephone (32) (2) 242 7284
Contact J. B. Boulton
Categories A1 B2

Jaeger Graphic Technology
20 Avenue des Desirs
B-1140 Brussels
Belgium
Telephone (32) (2) 7359551
Contact: M. Jaeger
Telex: 23027 JAEGER B
FAX: (32)(2) 733 10 35
Categories A7

HOLOGRAM PRODUCERS

A1= Commercial H-1
A2=Fine Art Originals
A3=Fine Art Ltd Editions
A4=Custom Presentation Support

A5=Holographic Stereograms
A6=Embossed/Hotstamped
A7=Silver Halide Transmission
A8=Silver Halide Reflection

A9= Custom Lighting Support
A10=Oversize/Large Format
A11=Pulse Holograms
A12=Dichromate

BRAZIL

Holografica
Rua California 978
Brooklyn, Sao Paulo
Brazil
Telephone (55) (11) 530 0169
Categories A1

CANADA

Claudette Abrams
22 Bayview Avenue
Wards Island, Toronto
Ontario, Canada
M5J IZI
Telephone (416) 533 4892
Categories: A2

Aptec Engineering Limited
4251 Steeles Avenue West
Downsview
Ontario, M3N IV7
Canada
Telephone (1) (416) 6619722
Categories: A1

Associates of Science and Technology
2450 Lancaster Road, Suite 36
Ottawa KlB 5N3 Canada
Telephone (1) (613) 521 2557
Contact: Dr. J.William McGowan
Categories: A8

Meryn Cadell
32 Augusta Ave.
Toronto M5T 2K8
Canada
Telephone (1) (416) 368 5593
Categories A1

Marie-Andree Cossette
1145 Avenue des Laurentides
Quebec City, Quebec GI5 3C2
Canada
Telephone (1) (418) 6872985
Categories A1

Melissa Crenshaw
2525 York
Vancouver, B.C.
Canada V6K IE4
Telephone (1) (604) 7341614
Categories: A1 A2

HOLOGRAM PRODUCERS

A1= Commercial H-1
A2=Fine Art Originals
A3=Fine Art Ltd Editions
A4=Custom Presentation
    Support

A5=Holographic Stereograms
A6=Embossed/Hotstamped
A7=Silver Halide Transmission
A8=Silver Halide Reflection

A9= Custom Lighting Support
A10=Oversize/Large Format
A11=Pulse Holograms
A12=Dichromate

Deep Space Holographics
1328 Dunsterville Ave.
Victoria, British Columbia
Canada V8Z 2Xl
Telephone (1)(604) 4794357
Categories A1

Sydney Dinsmore
101 MacDonell Avenue
Toronto, Ontario
Canada M6R 2A4
Telephone (1)(416) 536 2877
Categories: A1

Deborah Duston
90 Sherbrooke Avenue
Ottawa, Ontario
Canada K14 lR9
Telephone (1) (613) 722 9004
Categories A1

Georges M. Dyens
1293 Rue de la Visitation
Montreal, Quebec
Canada H2L 3B6
Telephone (1) (514) 5988860
Categories A1

Jacques Frigon
2375 Fullum
Montreal, Quebec
Canada H2K 3P3
Telephone (1) (514) 521 4270
Categories A1

Fringe Research Holographics
1179A King Street, West Suite 008
Toronto, Ontario
Canada M6K 3C5
Telephone (416) 5352323
Contact: Michael Sowdon, Dir.
Categories A2 A3 A7 A8 All Bl B5 C4 C8 C10

General Holographics, Inc.
P.O. Box 82247
Vancouver B.C., V5C 5P7
Canada
Telephone (1) (604) 299 6618
Contact. Bernd Simson
Categories Al A12

Lorna Heaton
3998 Rue de Bullion
Montreal, Quebec
Canada H2W 2E4
Telephone (1)(514) 8455403
Categories Al

Holocor IBF. Printing Inc.
95 des Sulpiciens
L'Epiphanie, Quebec
Canada JOK IJO
Telephone (514) 588 5524
Contact. Jean-Robert Bernier
FAX: (514) 588 4898
Categories A6

Holocrafts: Division of Canadian Holographic
Developments
Box 1035
Delta, B.C.
Canada, V4M 3T2
Telephone (1) (604) 9461926
Contact: Karoline Cullen
FAX: (604) 9461648
Categories Al AS A12

Holo-Dimensions Inc.
3577 Rue de Bullion
Montreal, Quebec
Canada H2X 3Al
Telephone (1) (514) 8454419
Categories Al

HOLOGRAM PRODUCERS

A1= Commercial H-1	**A5**=Holographic Stereograms	**A9**= Custom Lighting Support
A2=Fine Art Originals	**A6**=Embossed/Hotstamped	**A10**=Oversize/Large Format
A3=Fine Art Ltd Editions	**A7**=Silver Halide Transmission	**A11**=Pulse Holograms
A4=Custom Presentation Support	**A8**=Silver Halide Reflection	**A12**=Dichromate

Holo Laser Tech
7 Fraser Avenue #16
Toronto, Ontario
Canada, M6K 1Y7
Telephone (1) (416) 638-7991
Contact: Glen Stradz
Categories: A1 B3

Holomagic
Ruth Simkin, President
917 17th Avenue, S.W.
Calgary, Alberta
Canada T2T OA4
Telephone (1) (403) 229 0069
Categories: A1

Holomorph Visuals, Inc
273 de la Gauchetiere W.
Montreal, Quebec H27 1C7
Canada
Telephone. (1)(514) 8724530
Contact: Kenneth T. Chalk
Categories A1

Holospectra Inc.
CP8509
Ste-Foy, Quebec
Canada GIV 4N5
Telephone. (1) (418) 6515870
Categories: A1

HOLOGRAM PRODUCERS

A1= Commercial H-1
A2=Fine Art Originals
A3=Fine Art Ltd Editions
A4=Custom Presentation
    Support

A5=Holographic Stereograms
A6=Embossed/Hotstamped
A7=Silver Halide Transmission
A8=Silver Halide Reflection

A9= Custom Lighting Support
A10=Oversize/Large Format
A11=Pulse Holograms
A12=Dichromate

Holospectra Ine.
840, Ste-Therese
Quebec City, Quebec
Canada, GIN IS7
Telephone (418) 6833530
Contact: Alain Beauregard
FAX: (418) 6825594
Categories: Al A6 A 7 A8

IBOU.
CP214
Cap-de-Ia-Madeleine
Quebec, Canada G8T 7W2
Telephone (1) (819) 2955229
Categories: A1 A2 A3 A4 A7 A8 A9 B2 B3 B4 B5
B6 B8 B9

Anik Lafreniere
4060 Boulevard St. Laurent
Montreal, Quebec
Canada H2W 1Y9
Telephone (1)(514) 2869619
Categories: A1

Laser Holographies, Inc.
1179 King St. West
Unit III
Toronto, Ontario
Canada M6K 3C5
Telephone (1) (416) 531 4656
Contact: Charles Demicher
Telex. 7601314 (UI)
FAX: (1)(416) 5301594
Categories: A1 A6 B1 B3 B4 B5 B6 B7 B8 B9

Laser Innovations Inc.
25 Fisherville Road, Unit 804
Willowdale, Toronto
Ontario M2R 3B7
Canada
Telephone (1)(416) 8611747
Categories A1

Laval University
1145 Ave. Des Laurentides
Quebec City, Quebec
Canada GIS 3C2
Telephone (1) (418) 6872985
Contact Marie Cossette
Categories: Al A2 A3 B5 C3 C4 C5 C12

Les Productions Hololab!
Attn: Marie-Christiane Mathieu
3970, Boulevarde St. Laurent
Montreal, Quebec H2W IY3
Canada
Telephone (1) (514) 8494325
Categories A1

Light Construction, Inc.
2154 Dundas Street West
Suite #503
Toronto, Ontario
Canada M6R IX3
Telephone (1) (416) 5334692
FAX: (1)(416) 5330572
Categories: Al A2 A7 A8 A10 B5 C12

Jean-Pierre Marchand
C.P.214
Cap-de-Ia-Madeleine
Quebec
Canada, G8T 7W2
Telephone (1) (819) 2955229
Categories A2 A3 A7

MGD Modulations
1293 Rue de la Visitation
Montreal, Quebec
Canada
Telephone (1) (514) 5988860
Categories A1 A2

HOLOGRAM PRODUCERS

A1= Commercial H-1	**A5**=Holographic Stereograms	**A9**= Custom Lighting Support
A2=Fine Art Originals	**A6**=Embossed/Hotstamped	**A10**=Oversize/Large Format
A3=Fine Art Ltd Editions	**A7**=Silver Halide Transmission	**A11**=Pulse Holograms
A4=Custom Presentation Support	**A8**=Silver Halide Reflection	**A12**=Dichromate

RobertMyre
6090 Waverly
Montreal, Quebec
Canada H2X 2M
Telephone (1) (416) 5334692
Categories A1

Joanne Roy
3575 Rue de Bullion
Montreal, Quebec
Canada H2X 3A1
Telephone (1) (514) 8454419
Categories A1

Bernd Simson
Box 82247
Burnaby, B.C.
Canada V5C 5PT
Telephone (1) (604) 4356654
Categories: A1

Trilone Holographie Corp.
4200 Blvd. St Laurent
Montreal, Quebec
H2W 2R2 Canada
Telephone (1)(514) 8456992
Contact Gerard Allon
FAX: (1)(514) 8498706
Categories: Al A2 A3 A4 A7 A8 A9 A10 A11 B3 B5 B8 B9

DENMARK

Frithioff Johansen
Dantes Plads 4
DK 1556 Copenhagen V
Demnark
Telephone (45) (01) 150328
Phone. (45)(01) 103928
Categories A2 A3

FRANCE

ElF PRODUCTIONS
EIZYKMAN/FIHMAN
19 Rue Jean Jacques Rousseau
F, 75001 Paris, France
Telephone (33) (1)4236 0631
Contact. Claudine Eizykman
Categories A1

Jean Gilles
14 Rue des Quatre Vents
25000 Besancon, France
Telephone (33) (81) 503 139
Categories A1

Holographic Creations
Attn: J-C Raverat de Boisheu
26- Rue Daniel Stem
Paris 75015, France
Telephone (33) (14) 5788742
Categories A1

Hologram. Industries
22-24 rue des Oseraies
93100 Montreuil
France
Telephone (33) (14) 8700099
Contact. Hugues Souparis
FAX: (33)(14) 8701329
Categories A1

Holo-Laser
12, Rue De Vouille
75015 Paris, France
Telephone (33) (1) 45315275
Contact: Dr.J.L. Tribillon
Categories: A1 A2 A3 A4 A6 A7 A8 A10 C1 C10 C11

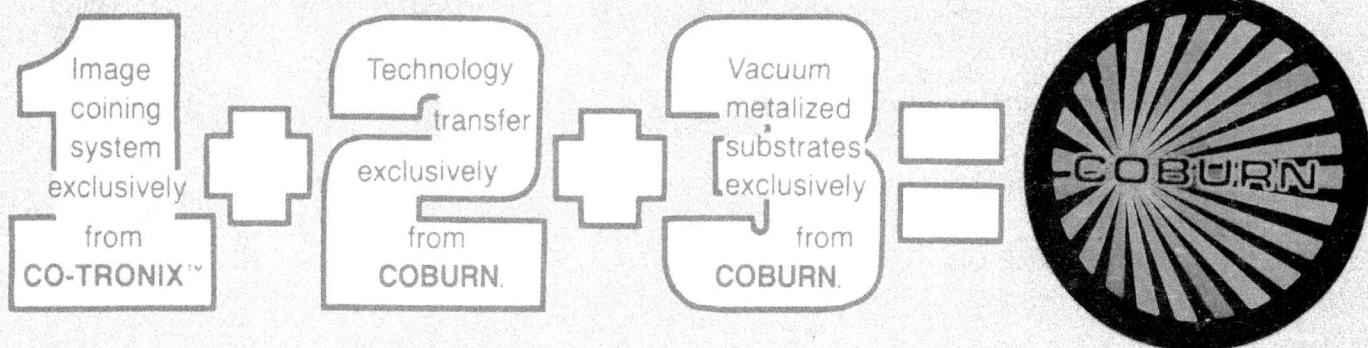

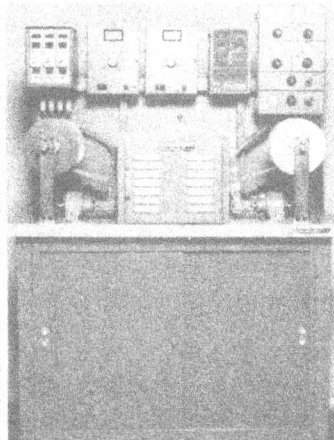

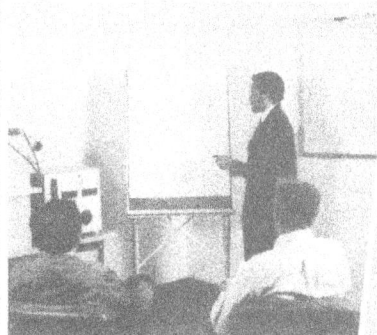

COBURN CORPORATION

Coburn Corporation lifts the vail of mystery by offering an intensified 3-day, one-on-one, hands-on course relating to the creation and production of 2-D holographic imagery. This program steps the participant through all phases of mastering techniques from concept and graphic preps to final mechanical laboratory procedures. All equipment and supplies required in the teaching-learning process are included. After successful completion of this course, individuals will possess the facility to routinely generate 2-D holographic masters for replication.

Prerequisites: Candidates should possess the ability to mentally visualize the end product including color separations and effect from line drawing; some knowledge of graphic/photographic procedures and mechanical lab operations helpful.

Course: H-1001; three days; instructed at Coburn's Lakewood (N.J.) Industrial Park Complex

For more information, contact Coburn Corporation at the address lower right.

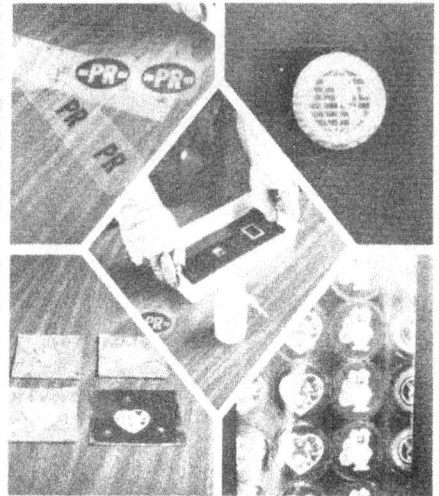

2-D HOLOGRAPHIC MASTERING

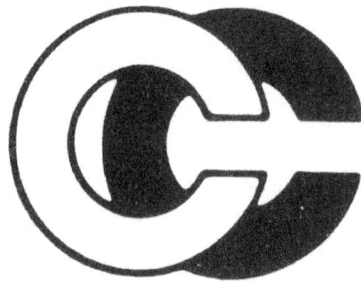

TECHNOLOGY TRANSFER

EMBOSSING SHIMS ELECTROFORMING (INCLUDING EQUIPMENT)

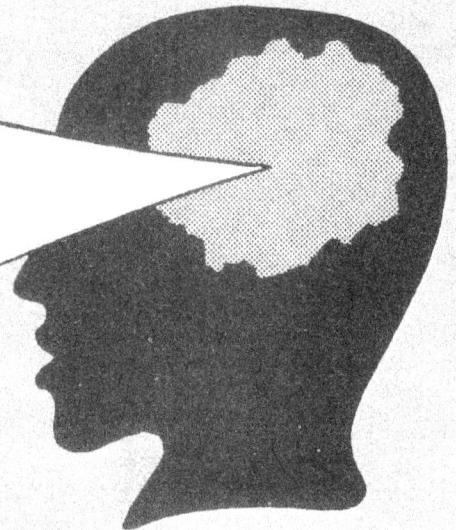

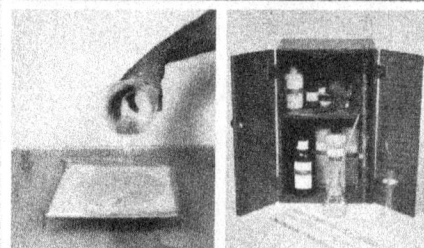

Electroforming of finished embossing plates to produce 2-D, 3-D holograms or other dimensional patterns and images has been regarded by many as a highly proprietary procedure. Coburn Corporation now offers this technology combined with all specialized equipment required to interested parties in our industry. Individuals enrolled will participate in a concentrated three day one-on-one hands-on program encompassing all facets of this technology, procedure and equipment familiarization/operation. After successful course completion, the individual will have acquired the skills to produce flawless metal embossing plates (embossing "shims") from glass or plastic masters plus a complete production facility including equipment plus start-up materials and chemistry.

For more information on 2-D holographic mastering or embossing plate electroforming; contact Coburn at the address below.

COBURN CORPORATION

1650 Corporate Road West,
Lakewood, NJ 08701 USA
Tel. (201) 367-5511 Tlx 13-2438
Fax (201) 367-2908 Cable: Unique

HOLOGRAM PRODUCERS

A1= Commercial H-1
A2=Fine Art Originals
A3=Fine Art Ltd Editions
A4=Custom Presentation
    Support

A5=Holographic Stereograms
A6=Embossed/Hotstamped
A7=Silver Halide Transmission
A8=Silver Halide Reflection

A9= Custom Lighting Support
A10=Oversize/Large Format
A11=Pulse Holograms
A12=Dichromate

Holo-Visual Concept 3D
25, Rue Cavendish
Paris, 75019, France
Telephone (33) (14) 200 2935
Contact Melanie Harris
Categories: A1

IDHOL
Boite Postale 7
F.89340 Saint-Agnan
France
Telephone (33) (16) 86961929
Contact: Jacques Bousigue
Telex. 699559 IDH03333
Categories A1 A2 A3 A4 A7 A8 A10

Optical Laboratory
3, Rue de Universite
67000 Strasbourg
France
Categories: A1

Yannis Palamas
1 Villa de Guelma
75018 Paris
France
Telephone (33) (14) 42556057
Categories: A1

Jacques Senechal
3 Rue St. Denis
92100 Boulogne - Billancourt
France
Categories: A1

J. Louis Tribillon
12 Rue de Vouille
Paris 75015
France
Telephone (33) (14) 531 5275
Categories A1

X-JAL
7 Rue de l'Universite
F-6000 Strasbourg
France
Telephone (33) (88) 352 132
Categories A1

HUNGARY

Artplay Holographic Studio
H-1191 Budapest
ADYENDRE ut 8.
Hungary
Telephone (36) (361) 270412
Contact Tibor Balogh
Categories: A1 A2 A3 A4 A6 A7 A8 A12

ISRAEL

Holofar Lab Israel
RH 2 November N14
Haifa
Israel
Telephone (972) (4) 332705
Categories A1

ITALY

Vittorio Alliata
SOI
Via Degli Eugenii 23
Rome 00178
Italy
Telephone (39) (6) 799 0452
Categories A1

Bimbolegge & Bimbogioca SRL
Via Borfuro 12
Bergamo 24100
Italy
Telephone (39) (35) 213015
Categories A1

HOLOGRAM PRODUCERS

A1= Commercial H-1	**A5**=Holographic Stereograms	**A9**= Custom Lighting Support
A2=Fine Art Originals	**A6**=Embossed/Hotstamped	**A10**=Oversize/Large Format
A3=Fine Art Ltd Editions	**A7**=Silver Halide Transmission	**A11**=Pulse Holograms
A4=Custom Presentation	**A8**=Silver Halide Reflection	**A12**=Dichromate
Support		

Holofur Lab (SRL)
Piazza Acilia No.3, Int. 3
Rome
Italy 00199
Telephone (39) (6) 8395517
Categories: A1

JAPAN

C.Itoh & Company
Central P.O. Box 136
Tokyo 100-91
Japan
Telephone (81)(3) 6392946
Categories A1

HoloARP
Kamakura Inc.
7-10-8 Ginza Chuo-Ku
Tokyo
Japan
Telephone (81)(03) 5748307
Contact: Yumiko Shiozaki
FAX: (81) (03) 5748377
Categories: Al A2 A3 A4 A6 B1 B3 B5 C3 C6 C9

Setsuko Ishii
1-23-26-404 Kohinata
Bunkyo-Ku
Tokyo
Japan
Telephone (81) (3) 9459017
Categories: A1

Yumiko Shiozaki
7-5-18-Ryoke Urawa
Saitama
Japan
Telephone (81) (0488) 31 7723
Contact: Yumiko Shiozaki
Categories: A2 B5 C3 C5 C7 C8 C9 C10

Shunsuke Mitamura
Institute of Art & Design
University of Tsukuba
Sakura-mura, Niihari-gun
Ibaraki-ken 305,Japan
Telephone (81) (298) 532833
Categories A1

LUXEMBOURG

David Dewar
39 Rue D 'Oradour
Luxembourg 2266
Luxembourg
Telephone (352) 444 437
Categories A1

MEXICO

Hologramas De Mexico
PINO 343, Local 3
06400 Mexico, D.F. Mexico
Telephone (52) (905) 5479046
Contact: Dan Lieberman
FAX (52) (905) 5474084
Categories: A1 A2 A4 A5 A6 A8 A9 A12 B2 B3 B6 C10

NETHERLANDS

Alexander Coblijn
P.O. Box2G
9700AA
Groningen
The Netherlands
Telephone (31)(50) 140417
Categories A1

HOLOGRAM PRODUCERS

A1= Commercial H-1   **A5**=Holographic Stereograms   **A9**= Custom Lighting Support
A2=Fine Art Originals   **A6**=Embossed/Hotstamped   **A10**=Oversize/Large Format
A3=Fine Art Ltd Editions   **A7**=Silver Halide Transmission   **A11**=Pulse Holograms
A4=Custom Presentation   **A8**=Silver Halide Reflection   **A12**=Dichromate
         Support

Dutch Holographic Laboratory
Kanaaldyk Noord 61
Eindhoven 5642JA
The Netherlands
Telephone (31) (40) 817250
Contact: Walter Spierings, Dir.
FAX: (31)(40) 814865
Categories: Al A3 A4 AS A6 A7 A8 A9 AI0 B3 Dl

Foundation Ideecentrum
P.O. Box 222
5600MK
Eindhoven
The Netherlands
Categories A1

Que Sera Sera
P.O. Box 29
9700 AA Groningen
The Netherlands
Telephone (31)(050) 140417
Contact: H. T. Vogd
FAX: (31) (050) 144142
Categories A12 B3 C4 D3 D4

Walter Spierings
Geldron pseweg 94B
Eindhoven 5611 SL
Netherlands
Telephone (31) (40) 115202
FAX: (31) (40) 814865
Categories: Al

White Tiger Holograms
Johannes Verhulststraat 45
1071 MS Amsterdam
The Netherlands
Telephone (31) (20) 797182
Contact: Neil Walker
FAX: (31)(20) 790896
Categories A1 AS A6 A12

SOUTH AFRICA

HoIotecCC
P.O. Box 5144
Brackengardens
145 2 Transvaal
South Africa
Telephone (27)(011) 864 1292
Contact Mandy Van Der Molen
Categories: Al A2 A3 A4 A7 A8 A9

SPAIN

Antoni Pinol Gullamon
Anselmo Clave 74
Olesa de Montserrat
08640 Barcelona, Spain
Telephone (34) (3) 7782299
Categories A1

Holotek
Julio Ruiz Garcia
Granda 47 Siero
Asturias
Spain
Telephone (34) (985) 793526
Categories: Al B 7

Tridimensionale Hologramas
Alberto, Alcocer, 38-2D
28016 Madrid
Spain
Contact: Daniel Weiss
Categories: Al

SWEDEN

Torgny E. Carlsson
Vickervagen 4
14569 Norsborg
Sweden
Telephone (46) (753) 85659
Categories A1

YOUR DICTIONARY ISN'T UP TO DATE ANYMORE

PLEASE ADD: DUTCH HOLOGRAPHIC LABORATORY

Since at least 1984 DHL has occupied a leading position in the world of holograms.
One reason is its superbly equiped laboratories in which, with the aid of high-tech laser systems, an unending sequence of holograms is being developed.

And not just 'ordinary' reflection and transmission holograms on glass and film. For instance, DHL makes its own photo resist masters for embossed holograms on stickers or hot foil. But also transmission holograms measuring a square meter. Or do you want reflection holograms measuring 50 x 60 cm? DHL makes them!

DHL first received large-scale recognition, nationally and internationally, with its microscope. Other creations of ours have also attracted much attention, and still do.

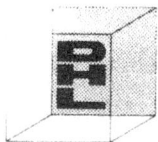

Thanks to the high-grade technical execution, but also, and in particular, to our surprising designs. Just name the application, and v come up with the ideas. Also if it is a matter of fitting into an existi product or design. DHL is right behind you.

Our services are much in demand by industrial giants such as Philips, Toshiba, Polygram, Shell, AT & T, ASEA and Akzo Chemi And now, we hope, by you.

The first step is to fill out the form below and send it to us. You v then receive our full-color brochure containing more specific information.

DUTCH HOLOGRAPHIC LABORATORY

Kanaaldijk Noord 61 5642 JA Eindhoven
The Netherlands
Telephone 040-81 7250 Telex 59032 notel nl.

QUALITY IN THREE DIMENSIONS

I'd like to see what DHL is capable of.
So send me your detailed full-color brochure.

Name:

Company:

Address:

Zipcode:

Telephone:

Send this coupon (or a copy) in a stamped envelope to:
DHL, Kanaaldijk Noord 61, 5642 JA Eindhoven, The Netherlands.

HOLOGRAM PRODUCERS

A1= Commercial H-1
A2=Fine Art Originals
A3=Fine Art Ltd Editions
A4=Custom Presentation
    Support

A5=Holographic Stereograms
A6=Embossed/Hotstamped
A7=Silver Halide Transmission
A8=Silver Halide Reflection

A9= Custom Lighting Support
A10=Oversize/Large Format
A11=Pulse Holograms
A12=Dichromate

EAJonsson
Laserlabbet
Box 521
SE581 06
Linkoping, Sweden
Telephone (46)(13) 123377
Categories: A1

Lasergruppen Holovision AB
Osthammavsgatan 69
S-11528 Stockholm
Sweden
Telephone (46)(8) 639908
Contact: Jonny Gustavsen
Categories: A1 A2 A3 A 7 A8 A10 A11

MarwellAB
Kyrkbacken 27
S-171 50 Solna
Sweden
Telephone (46) (8) 838 261
Categories: A1

SWITZERLAND

Pascal Barre
4 Place Grenus
1201 Geneva
Switzerland
Telephone (41) (22) 325 191
Categories A1

Laserart Ford AG
Rebenstrasse 20
CH 4125 Riehen/Basel
Switzerland
Telephone (41)(61) 672 343
Categories A1 A8 B2 B3

Carl Fredrick Reutersward
6 Rue Montolieu
1030 Bussigny/Lausanne
Switzerland
Telephone (41)(21) 890514
Categories: A1

TAIWAN

Tjing Ling Industrial Research
130 Keelung Road, Section III
Taipei, Taiwan
Telephone (86) (2) 704 1856
Categories: A2

UNITED KINGDOM

Advanced Holographics, Ltd.
243 Lower Mortlake Rd.,Unit 11
Richmond, Surrey TW9 2LL
England, United Kingdom
Telephone (44) (1) 9486811
Contact. Laurence Holden
FAX: (44) (1) 9488948
Categories: Al A2 A3 A4 A6 A 7 A8 A9 A10

A.H. Prismatic, Inc.
New England House
New England Street, Brighton
BNl 4GH England, United Kingdom
Telephone (44) (273) 686966
Contact: Simon Lee
FAX: (44) (273) 676692
Categories A8 B3

Matt Andrews
7 A Brunswick Park
Camberwell, London
SE5 7RH, England
United Kingdom
Telephone (44) (1) 7031254
Categories A1

HOLOGRAM PRODUCERS

A1= Commercial H-1
A2=Fine Art Originals
A3=Fine Art Ltd Editions
A4=Custom Presentation
        Support

A5=Holographic Stereograms
A6=Embossed/Hotstamped
A7=Silver Halide Transmission
A8=Silver Halide Reflection

A9= Custom Lighting Support
A10=Oversize/Large Format
A11=Pulse Holograms
A12=Dichromate

Applied Holographies
Braxted Park, Witham
Essex, CM8 3XB
England
United Kingdom
Telephone (44)(621) 893030
Categories A7 A8 B3

Ascot Laser Picture Studio
27 Upper Village Road
Sunninghill, Ascot
Berkshire SL5 7AJ
England, United Kingdom
Telephone (44)(0990) 21789
Contact: Mr. Brodel
Categories: A2 A3 A7 A8 C4 C9 C12

Barr & Stroud, Ltd.
Caxton Street
Anniesland
Glasgow G13 1HZ
Scotland, United Kingdom
Telephone (44)(41) 954 960 1
Categories A1

Margaret Benyon
Holography Studio
40 Springdale Avenue
Broadstone, Dorset
BH18 9EU, England
Telephone (0202) 698067
Contact: Margaret Benyon
Categories: A2 A3 A7 A8 B5 C4

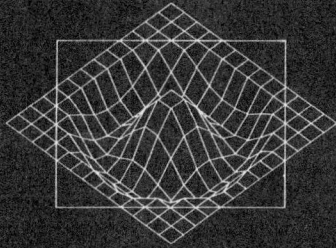

HOLOGRAM PRODUCERS

A1= Commercial H-1
A2=Fine Art Originals
A3=Fine Art Ltd Editions
A4=Custom Presentation
    Support

A5=Holographic Stereograms
A6=Embossed/Hotstamped
A7=Silver Halide Transmission
A8=Silver Halide Reflection

A9= Custom Lighting Support
A10=Oversize/Large Format
A11=Pulse Holograms
A12=Dichromate

Patrick Boyd
18 Whiteley Road
London SEI9IJT
England
United Kingdom
Telephone (44) (1) 6704160
Categories: A1 A2 A5 A11

Brodel Holograms
15 School Road
Sunninghill, Ascot
Berkshire, England
SL57AE
United Kingdom
Telephone (44) (990) 21789
Categories: A1 A2 A5 A11

Peter Buxton
56 Courtfield Gardens
London, SWS ONF
England
United Kingdom
Telephone (44)(1) 3706996
Categories. A1

Angela Coombes
Dovecote Studio
Witham Friary
Frome, Somerset
England
United Kingdom
Telephone (44) (74) 985 691
Categories: A1 A10

Susan Ann Cowles
123 Grove Green Road
London Ell 4ED
England
United Kingdom
Categories: A1 A3

Darkroom Eight Ltd.
Unit 8 - Impress house
Vale Grove, Acton
London W3 7QH
United Kingdom
Telephone (44)(1) 7492218
Categories A1 C4

Diversified Optical Ltd.
39 Cambridge Road
Kesgrave, Ipswich
Suffolk IP5 7EW
United Kingdom
Telephone (44) (473) 625 468
Categories A1

Farsound Ltd.
23 Englefield Road
London, NI 4EJ
England, United Kingdom
Telephone (44)(1) 2546553
Categories A1

Nick Hardy
Unit 4, Technorth
7 Harrogate Road
Leeds LS7 3NB
England, United Kingdom
Telephone (44) (943) 628687
Categories A1

Ken Harris
41 Withdean Road
Preston Park
Brighton BNI 5FE
England, United Kingdom
Categories A1

HOLOGRAM PRODUCERS

A1= Commercial H-1
A2=Fine Art Originals
A3=Fine Art Ltd Editions
A4=Custom Presentation Support

A5=Holographic Stereograms
A6=Embossed/Hotstamped
A7=Silver Halide Transmission
A8=Silver Halide Reflection

A9= Custom Lighting Support
A10=Oversize/Large Format
A11=Pulse Holograms
A12=Dichromate

Anne Hickmott
11 Castlenau
London SWI3 9RP
England,United Kingdom
Telephone (44)(1) 7481031
Categories A1

Holofax Limited
Netherwood Road
Rotherwas Industrial Estate
Hereford HR2 6JZ
England, United Kingdom
Telephone (44)(432) 278400
Categories: A8 D6

Hologram One
39 Pyrcoft Road
Chertsey, Surrey
England KTI6 9HT
United Kingdom
Telephone (44) (9328) 64899
Categories A1 A8 B3

Holographies (UK) Ltd.
32 Lexington Street
London WIR 3HR
England, United Kingdom
Telephone (44) (1) 4378992
Contact Jon Vogel
FAX: (44)(1) 4940386
Categories: A1 A4 A6 A7 A8 A9 A10 A11 A12

Holoscan Ltd
The Laboratories, East
Tytherley Road, Loekerley
Hampshire, S051 OJT
England, United Kingdom
Telephone (44) (0794) 41229
Contact: Dr.John M. Webster
FAX: (44) (0794) 41264
Categories A1 A2 A4 A7 A8 A9 A10 A11

HoloteePLC
7 Cameron Road
Seven Kings, Ilford
Essex, 1 G3 BLG
England, United Kingdom
Telephone (44) (1) 5978004
Categories A1 B6 B7 B8

Holovision Ltd.
P.O. Box 555
London, W9 1 YG
England, United Kingdom
Telephone (44)(1) 289 9969
Categories A1

Anthony Hopkins
12 New StJohns Road
St. Helier,Jersey
Channel Islands, England
Telephone (44) (534) 30614
Categories A1

K.C. Brown Holographies
22 St Augustine's Road
Camden Town
London NWI 9RN
England, United Kingdom
Telephone (44) (1) 4822833
Categories A1 A11

HOLOGRAM PRODUCERS

A1= Commercial H-1	**A5**=Holographic Stereograms	**A9**= Custom Lighting Support
A2=Fine Art Originals	**A6**=Embossed/Hotstamped	**A10**=Oversize/Large Format
A3=Fine Art Ltd Editions	**A7**=Silver Halide Transmission	**A11**=Pulse Holograms
A4=Custom Presentation Support	**A8**=Silver Halide Reflection	**A12**=Dichromate

Andrew Laczynski
22 Woodstock Ave.
London WI3 9UG
England, United Kingdom
Telephone (44) (1) 840 5101
Categories: A1 B6

Chris Lambert
47 Alpine Street
Reading, Berkshire RGI 2PY
England, United Kingdom
Telephone (44) (734) 589 026
Categories A1

Laser Lines Ltd.
19 West Bar
Banbury, Oxon
OXI695A, England
Telephone (44) (295) 57581
Categories A1

Laza Holograms
47 Alpine Street
Reading, Berkshire
England RGI 2PY
United Kingdom
Telephone (44)(0734) 589026
Contact: Chris Lambert
Categories: A1 A2 AS A7 A8 A10 A11

Chris Levine
32 Lexington Street
London WIR3HR
England, United Kingdom
Telephone (44) (1) 437 8992
Categories: A1

Light Fantastic Ltd.
Trocadero
13 Coventry Street
London WIV 7FE England
Telephone (44)(1) 7344516
Categories: A1 A2 A3 A4 A5 A6 A7 A8 A9 A10 A11
A12 B1 B2 B3 B4 B5 B6 B7 B8 B9 C1 C8 C10

Light Fantastic PLC
48 South Row, The Market
Covent Garden
London WC2E 8HN
England
Telephone (44)(1) 8366423
Contact Roger Knight
FAX: (44)(1) 8364292
Categories: A1 A2 A3 A4 A5 A6 A 7 A8 A9 A10 A11
A12 B1 B2 B3 B4 B5 B6 B7 B8 B9 C1 C8 C10

Light Impressions Europe, Ltd
Attn: Kenneth Harris
12 Mole Business Park, Off Sta
Leatherhead, Surrey KT22 7AQ
England, United Kingdom
Telephone (44) (372) 386 677
Categories: A1 A6 B3 B5

Lightworks
8IA Hatton Square
16/16A Baldwins Gardens
London ECIN 7RJ England
Telephone (44)(1) 4300028
Categories A1

Liliane Lijn
99 Camden Mews
London EC2A 2AA
England, United Kingdom
Telephone (44) (1) 4858524
Categories: A1

Andrew Logan
15 Appold Street
London EC2A 2AA
England, United Kingdom
Telephone (44) (1) 2471677
Categories A1

HOLOGRAM PRODUCERS

A1= Commercial H-1
A2=Fine Art Originals
A3=Fine Art Ltd Editions
A4=Custom Presentation
    Support

A5=Holographic Stereograms
A6=Embossed/Hotstamped
A7=Silver Halide Transmission
A8=Silver Halide Reflection

A9= Custom Lighting Support
A10=Oversize/Large Format
A11=Pulse Holograms
A12=Dichromate

Michael Medora
7 Barclay Road
LondonSW6
England, United Kingdom
Telephone (44) (1) 736 8639
Categories A1

Mirage Holograms Ltd.
Unit 2 Brook Lane Business Cen
Brook Lane North, Brentford
Middlesex TW8 OPP,
England, United Kingdom
Telephone (44)(1) 568 2454
Categories Al A 7 A8

Rohit Mistry
21 Edgwarebury Gardens
Edgware,
Middlesex HAS 811
England, United Kingdom
Telephone (44) (923) 246 760
Categories A1

Munday Spatial Imaging
39 Pyrcroft Road
Chertsey, Surrey
KTI6 9HT, England, U.K.
Telephone (44) (0932) 564 899
Categories: A1 A2 A3 A4 A7 A8 A11 B1 B2 B3 B4
B5

Newbold Wells Company
33 Paul Street
London EC2A 4JU
England, United Kingdom
Telephone (44) (1) 638 1471
Categories A1 A6 A7 A8

Paul Newman
Sheerwater Lodge, Sheerwater R
Woodham, Weybridge
Surrey KT15 3QL
England, United Kingdom
Telephone (44) (1) 09323
FAX: (44)(1) 42396
Categories A2 A3

OP-Graphics (Holography) Ltd.
Unit 4 : Technorth
7 Harrogate Road
Leeds LS7 3NB
England, United Kingdom
Telephone (44) (0532) 628687
Contact: V. Love, N. Hardy
FAX: (44) (0943) 467881
Categories: A1 A2 A3 A4 A 7 A8 A9 A10 A11

Optical Engineering Group
Dept. of Mechanical Engineerin
Loughborough University of Technology
Loughborough, Leicestershire
England, United Kingdom
Telephone (44) (0509) 223222
Contact: John Tyror
FAX: (44) (0509) 231 983
Categories: A1 A2 A4 A5 A6 A7 A8 A9 A11 A12 C3
C4 C8

Optical Surfaces Ltd.
Godstone Road
Kenley, Surrey
CR2 5AA, England
United Kingdom
Telephone (44) (1) 668 6126
Categories A1

Optical Works Ltd.
32 The Mall
Ealing, London
W5 3TJ, England
United Kingdom
Telephone (44) (1) 5675678
Categories A1

Oriel Scientific Ltd.
P.O. Box 31
Leatherhead, Surrey
KT22 7AU, England
United Kingdom
Categories A1

HOLOGRAM PRODUCERS

A1= Commercial H-1	**A5**=Holographic Stereograms	**A9**= Custom Lighting Support
A2=Fine Art Originals	**A6**=Embossed/Hotstamped	**A10**=Oversize/Large Format
A3=Fine Art Ltd Editions	**A7**=Silver Halide Transmission	**A11**=Pulse Holograms
A4=Custom Presentation Support	**A8**=Silver Halide Reflection	**A12**=Dichromate

Oxford Holographics
71 High Street
Oxford OXI 4BA
England
United Kingdom
Telephone (44)(865) 250505
Categories A1

Oxford Scientific
113 Lavender Hill
Tonbridge, Kent
TN92AY
England, United Kingdom
Telephone (44) (732) 364 002
Categories Al

Caroline Palmer
49 Upper Woodford
Salisbury, Wiltshire SP4 6NU
England
United Kingdom
Telephone (44)(72) 273471
Categories: A1

Andrew Pepper
22 Haldane Road
LondonE63]
England, United Kingdom
Telephone (44) (1) 4711609
FAX: (44)(1) 3181439
Categories: A2 A3 A8 C6 C7 C9

Perception Holography
Thornton Marketing Ltd.
Aketon Close, Haggs Lane
Follifoot, Harrogate
North Yorks.,England HG3 lA2
Telephone (44) (93) 782323
Contact Mike Burridge
Categories A1 A7 A8 A11 C4 C9

Pilkington P .E. Ltd.
Glascoed Road, St. Asaph
Clwyd, England
LL17011
United Kingdom
Telephone (44) (745) 583 301
Categories A1

David Pizzanelli
4 Macaulay Road
London, SW4 OQX
England
United Kingdom
Telephone (44) (1) 6271140
Categories: Al A2 A3 A4 A6 A7 A8 A9 B7 C9

Raven Holo Ltd.
Old Saw Mills, Nyewood
Near Petersfield, Hampshire
GU3 I5HX, England
United Kingdom
Categories A1

Richmond Holographic Studios
6 Marlborough Road
Richmond, Surrey
England, TWIO 6JR
United Kingdom
Telephone (44) (1) 9405525
Contact: E.M.Orr, D. Traynor
Telex. 932905 LARCH G
FAX: (44)(1) 9486214
Categories: A1 A2 A3 A4 A5 A7 A8 A9 A10 A11

Rolfin Ltd.
Winslade House
Egham Hill, Egham Surrey
TW20 OAZ, England
Telephone (44)(7843) 7541
Categories A1

HOLOGRAM PRODUCERS

A1= Commercial H-1	**A5**=Holographic Stereograms	**A9**= Custom Lighting Support
A2=Fine Art Originals	**A6**=Embossed/Hotstamped	**A10**=Oversize/Large Format
A3=Fine Art Ltd Editions	**A7**=Silver Halide Transmission	**A11**=Pulse Holograms
A4=Custom Presentation Support	**A8**=Silver Halide Reflection	**A12**=Dichromate

Rosewell Ltd.
Blacknest Estate
Bentley, Alton
Hants., England
Telephone (44) (44) 42023605
Contact: Neville McGeorge
FAX: (44)(44) 42022517
Categories A1 A6

Scope Optics Ltd.
Unit 6 Alston Works
Alston Road, Barnet
Hertfordshire EN5 4EL, England
Telephone (44)(1) 441 2283
Categories A1

See 3 Holograms Ltd.
4 Macaulay Road
London, SW4 OQX, England
Telephone (44) (1) 6227729
Contact: D. Pizzanelli, J.Ross
Telex. 8951182 GECOMSG
FAX: (44)(1) 6225308
Categories: A1 A6

Jacques Senechal
90 Temple Fortune Lane
London NWII 7TX
England, United Kingdom
Telephone (44) (1) 4585825
Categories: A1 A10

Rick Silberman
2 Foxes Lane
Mouse Hole, Cornwall
England, United Kingdom
Telephone (44) (736) 731067
Categories A1

SPECAC Limited
6A River House
Lagoon Road, St. Mary Cray
Orpington, Kent
BR5 3QX, England, U.K.
Telephone (44) (689) 7317
Categories A1

Spectrolab Limited.
P.O. Box 25
Newbury, Berkshire
RGI3 2AD, England
United Kingdom
Telephone (44) (635) 248 080
Categories A1

TechNorth
Unit 4
7 Harrogate Road
Leeds LS7 3NB
England
Telephone (44) (532) 628687
Categories A1

Third Dimension Ltd.
4 Wellington Park Estate
Waterloo Road
LondonNW2~
England, United Kingdom
Telephone (44) (1) 208 0788
Categories A1

T ouchwood Holographic Studio
50 Sugworth Lane
Radley, Abingdon
Oxon, OXI4 2HY
England, United Kingdom
Telephone (44) (865) 735874
Categories A1

HOLOGRAM PRODUCERS

A1= Commercial H-1	**A5**=Holographic Stereograms	**A9**= Custom Lighting Support
A2=Fine Art Originals	**A6**=Embossed/Hotstamped	**A10**=Oversize/Large Format
A3=Fine Art Ltd Editions	**A7**=Silver Halide Transmission	**A11**=Pulse Holograms
A4=Custom Presentation Support	**A8**=Silver Halide Reflection	**A12**=Dichromate

Graham Tunnadine
46 Calthrope Street
London, WCl, England
Telephone (44)(1) 2781572
Categories A1

Michael Waller-Bridge
40 St Peter's Square
London W69AA
England, United Kingdom
Telephone (44)(1) 748 3851
Categories A1

Wenyon & Gamble
8 Berry Street
London ECIV OAU
England, United Kingdom
Telephone (44)(1) 2511797
Categories A2 C6

Wotan Lamps Ltd
1 Gresham Way
Dumsford Road
London, SW19 England
Telephone (44)(1) 9471261
Categories: A1

UNITED STATES OF AMERICA

Abbott Laboratories
Department 93F (Building AP-9)
Routes 43 and 137
Abbott Park, IL 60064
Telephone (1) (312) 9374117
Contact: Dr. Gerald Cohn
Categories A2

AD 2000
946 State Street
New Haven, CT
06511
Telephone (1) (800) 3344633
Contact: Jeffrey Levine
Categories: Al A4 A6 A7 A8 A9 A12 B7 B8 B9

Advanced Dimensional Displays
16742 Stagg Street- Suite 102
VanNuys,CA
91400
Telephone (1) (818) 7856563
Contact Steve Booth
Telex. 9103332519
Categories A5

Advanced Environmental Research
Route 1, Box 1830
Woolwich, ME
04579
Telephone (1)(207) 4436587
Contact: KKing, R Ian, S. Weber
Categories A6 A7

A.H. Prismatic Inc.
285 West Broadway
New York, NY
10013
Telephone (1) (212)219 0440
Contact Andrew Meehan
FAX: (1)(212) 2190443
Categories: A8 B1 B3 B4

Aites Lightworks
2148 North 86th Street
Seattle, WA 98103
Telephone (1) (206) 5265752
Contact: Edward Aites
Categories: A2 A3 A7 A8 B2 B5

"Alexander"
1323 14th Street, Apt. L
Santa Monica, CA 90404
Telephone (1) (213) 3939846
FAX: (1) (213) 4515921
Categories A2 A3 A7 A8 AI0 All

HOLOGRAM PRODUCERS

A1= Commercial H-1
A2=Fine Art Originals
A3=Fine Art Ltd Editions
A4=Custom Presentation Support

A5=Holographic Stereograms
A6=Embossed/Hotstamped
A7=Silver Halide Transmission
A8=Silver Halide Reflection

A9= Custom Lighting Support
A10=Oversize/Large Format
A11=Pulse Holograms
A12=Dichromate

American Bank Note Holographics
Woodfield Park Office Plaza
999 Plaza Drive, Suite 400
Shaumburg, IL 60173
USA
Telephone (1)(312) 3104494
Categories A6

Amherst Media
418 HomecrestDrive
Amherst, NY
14226
Telephone (1)(716) 8341480
Categories A1

Anait Studio
1685 Fernald Point Lane
Santa Barbara, CA
93108
Telephone (1)(805) 9695666
Contact:. Anait
Categories: A2 A3 All B5 C4 C12

Applied Holographics, Corp.
1530 Progress Road
P.O. Box 8300
Fort Wayne, IN
46898
Telephone (1) (219) 4846081
Contact. Richard W. Gipp
FAX:. (1)(219) 4848611
Categories: A1 A4 A6 A7 A8 A11 B8 B9

Armstrong World Industries
P.O. Box 3511
Lancaster, PA 17604
Telephone (1) (717) 3970611
Contact Larrimore B. Emmons
Categories A1

Artigliography
7130 Mohawk West Drive
Indianapolis, IN 46236
Telephone (1)(317) 8230069
Contact Kerry S. Brown
Categories: A2 A 7 A8 B3 B5 C3 C6 C8 C11

ArtKitek
122 Myrtle Avenue
Cotati, CA
94928
Telephone (1) (707) 664 2330
Categories A1

HaleAust
521 East 12th Street
New York, NY 10009
Telephone (1) (212)982 3713
Categories A2

Bliss, Barefoot and Associates
500 Fifth Avenue
New York, NY
10110
Telephone (1) (212) 8401661
Contact:. Paul D. Barefoot
FAX:. (1) (212) 8401663
Categories: A1 B 7

Joseph Belk
973 Page Street Studio
San Francisco, CA
94117
Telephone (1) (415) 431 9581
Categories A2

Rudie Berkhout
223 West 21st Street, Apt. B
New York, NY
10011
Telephone (1) (212) 2557569
Categories A2 A3 B5

HOLOGRAM PRODUCERS

A1= Commercial H-1
A2=Fine Art Originals
A3=Fine Art Ltd Editions
A4=Custom Presentation Support

A5=Holographic Stereograms
A6=Embossed/Hotstamped
A7=Silver Halide Transmission
A8=Silver Halide Reflection

A9= Custom Lighting Support
A10=Oversize/Large Format
A11=Pulse Holograms
A12=Dichromate

Roberta Booth Studio
Attn: Roberta Owen Booth
5326 Sunset Boulevard
Los Angeles, CA
90027
Telephone (1) (213) 466 5767
Categories A1

Roy Bradshaw
Reel Image
P.O. Box 566
Pacifica, CA 94044
Telephone (1) (415) 3558897
Categories A1

Brookhaven National Laboratory
Attn: Malcolm Richard Howells
Building Side
Upton, NY 11973
Telephone (1)(516) 2823758
Categories A1

Richard Bruck Holography
3312 West Belle Plaine #2
Chicago, IL
60618
Telephone (1)(312) 2679288
Categories A1 A2 A3 A4 A7 A8

Frank Bunts
15 West 24th Street
New York, NY
10010
Telephone (1) (212) 929 7938
Categories A2

Joseph Bums,Jr.
35'7 Pacific Street
Brooklyn, NY
11217
Telephone (1) (718) 6245966
Categories A2

Edward Bush
403 Highland Ave. No. 12F
Clifton, NJ
07043
Telephone (1) (201)4731280
Categories A2

Kenneth G. Byrne
177 East 4th Street
Brooklyn, NY
11218
Telephone (1) (718) 6334195
Categories A2

Cambridge Stereographics Group
P.O. Box 159
Kendall Square Station
Cambridge, MA
02142
Telephone (1)(617) 2530632
Categories A5

David Carlton
8934 Tarragon Court
Manassas, VA
22110
Telephone (1)(703) 361 9443
Categories A2

Casdin-Silver Holography
51 Melcher Street
Studio 501
Boston, MA
02210
Telephone (1)(617)4234717
Contact Harriet Casdin-Silver
Categories: Al A2 A3 A4 A5 A7 A8 A9 A10 A11 B2
B3 B5 B6 B8 B9 C2 C5 C6 C7 C9 C10 C11

HOLOGRAPHIC APPLICATIONS

"Usually there are many paths to a solution, and your holographic concept is no exception. Frequently the decisions made during the design stage can keep exciting options open, minimize risks and make the most effective use of this state of the art technology. Careful analysis of your application needs will result in the selection of the most appropriate holographic technique, material and production facility. Then you can comfortably watch your ideas blossom. Holography can be a great deal of fun; with our help it will be!"

Suzanne St.Cyr
Technical Director

HOLOGRAM PRODUCERS

A1= Commercial H-1
A2=Fine Art Originals
A3=Fine Art Ltd Editions
A4=Custom Presentation
      Support

A5=Holographic Stereograms
A6=Embossed/Hotstamped
A7=Silver Halide Transmission
A8=Silver Halide Reflection

A9= Custom Lighting Support
A10=Oversize/Large Format
A11=Pulse Holograms
A12=Dichromate

Casdin-Silver Holography
99 Pond Avenue
Suite 0403
Brookline, MA
02146
Telephone (1)(617) 7396869
Contact: Harriet Casdin-Silver
Categories: A1 A2 A3 A4 AS A7 A8 A9 A10 A11 B2
B3 B5 B6 B8 B9 C2 C5 C6 C7 C8 C10 C11 C12

Cherry Optical Company
2047 Blucher Valley Road
Sebastopol, CA
95472
Telephone (1) (707) 8237171
Contact: G.Cherry /N.Gorglione
Categories: A1 A2 A3 A4 A7 A8 A9 A10 B3 B4 B7
C2 C3 C6 C10

Coburn Corporation
1650 Corporate Road West
Lakewood, NJ
08701
Telephone (1) (201)367 5511
Contact: Joseph Coburn III
Telex 1324 38
FAX: (1) (201) 3672908
Categories: A6 B3

Arlie Conner
1514 South East Salmon
Portland, OR
97214
Telephone (1) (503) 2390545
Categories A2

Betsy Connors
12 Sunset Road
Somerville, MA
02144
USA
Telephone (1)(617) 6230578
Categories A2

Corion Corp.
73Jeffrey Ave.
Holliston, MA
01746
USA
Telephone (1) (617) 429 5065
Categories Al

Michael E. Crawford
4142 Bellefontaine
Houston, TX
77025-1105
USA
Telephone (1) (713) 6673375
Categories A1

Crown Roll Leaf, Inc.
91 Illinois Ave.
Paterson, N J.
07503
Telephone (1)(201) 7424000
Categories A6

Michael Croydon
950 Benson Lane
Green Oaks, IL
60048
Telephone (1) (312) 3628248
Categories A2

Thomas J. Cvetkovich
573 South Schenley
Youngstown, OH
44509
Telephone (1) (216) 799 0323
Categories A2

Frank Davis
3202 Argonne
Houston, TX 77098
Telephone (1) (713) 526 0006
Categories: A2

"Bouquet" by Nancy Gorglione, 1/1 Multicolor Reflection Hologram Composite

CHERRY OPTICAL HOLOGRAPHY
FINE ART HOLOGRAPHY BY
NANCY GORGLIONE

- One-of-a-Kind Reflection Hologram Composites
- Large Format Transmission Hologram Sculpture & Murals
- Architectural Installations and Lobby Art
- Stage Sets and Laser Lighting Effects
- Curatorial and Catalog Services

CHERRY OPTICAL HOLOGRAPHY
COMMERCIAL HOLOGRAPHY BY
GREGORY W. CHERRY

- Mastering and Origination for Reflection
  and Transmission Holograms
- Trade Show and Product Display Holograms
- Stock Copies – Best Sellers Worldwide. Our Images Include:
  - TELESCOPE – BINOCULARS – DONATIONS

You've seen our holograms at every major holography gallery and exhibit Worldwide

FINEST QUALITY AT THE MOST REASONABLE PRICES

CHERRY OPTICAL COMPANY
2047 BLUCHER VLY. RD.   SEBASTOPOL, CA 95472
(707) 823-7171

HOLOGRAM PRODUCERS

A1= Commercial H-1
A2=Fine Art Originals
A3=Fine Art Ltd Editions
A4=Custom Presentation
Support

A5=Holographic Stereograms
A6=Embossed/Hotstamped
A7=Silver Halide Transmission
A8=Silver Halide Reflection

A9= Custom Lighting Support
A10=Oversize/Large Format
A11=Pulse Holograms
A12=Dichromate

Dazzle Enterprises, Inc.
425 Southlake Blvd #Bl
Richmond, VA
23236, USA
Telephone (1)(804) 3795500
Contact: AJ.Languedoc, Pres.
FAX: (1) (804) 379 6328
Categories: A6

Frank DeFreitas
P.O. Box 9035
Allentown, PA 18105
Telephone (1) (215) 4348236
Categories A2

Julienne B. Derichs
5925 West Carol Avenue
Morton Grove, IL 60053
Categories: A2

Vincent DiBiase
P.O. Box 111
Lagunitas, CA 94938
Telephone (1) (415) 4544862
Categories A2

Edward Dietrich
2036 West Haddon
Chicago, IL, 60622
Telephone (1)(312) 2920770
Categories A2 C3 C5 C6 C8 C9

Diffraction Company, Inc.
P.O.Box 151
Riderwood, MD
21139
Telephone (1) (301) 6661144
Contact: Christopher Wynd
FAX: (1)(301) 472 4911
Categories: A6

Dimensional Imaging Technology
439 Walsingham Court
Dayton,OH
45429
Telephone (1) (513) 434 5818
Categories A2

Dimension Research
P.O. Box 2132
Southfield, MI
48037
Telephone (1) (313) 3550412
Contact:. Lee Lacey
FAX: (1)(313) 3550437
Categories: A1 A2 A3 A4 A5 A6 A7 A12

Douglass Associates Studios
3 Cove of Cork Lane
Annapolis, MD
21401
Categories Al

John Drevenak
10708 Lilac Avenue
St. Louis, MO
63137
Telephone (1)(314) 8694058
Categories A2

Philip Dubav
5401 Bee Cave Road
Austin, TX
78746
U&\
Telephone (512)327-5961
Categories A2

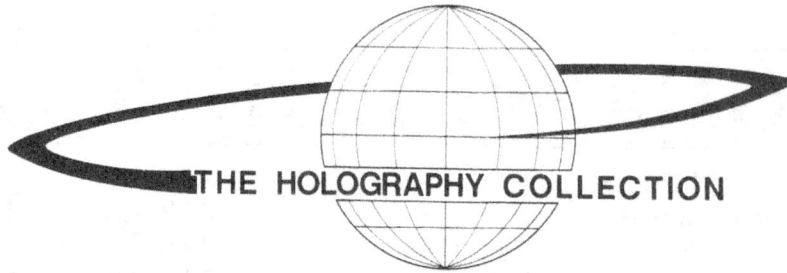

HOLOGRAM PRODUCERS

A1= Commercial H-1
A2=Fine Art Originals
A3=Fine Art Ltd Editions
A4=Custom Presentation
          Support

A5=Holographic Stereograms
A6=Embossed/Hotstamped
A7=Silver Halide Transmission
A8=Silver Halide Reflection

A9= Custom Lighting Support
A10=Oversize/Large Format
A11=Pulse Holograms
A12=Dichromate

DZCompany
181 Mayhew Way Suite E
P.O. Box 5047-T
Walnut Creek, CA
94596
Telephone (1)(415) 9354656
Contact: Dan Cifelli
FAX: (1)(415) 9354660
Categories: A1 A4 A6 B3 B6 B7 B8

Eastman Kodak Company
Wayne Shaffer
1669 Lake Avenue, B-23, Fl. 3
Rochester, NY
14650
Telephone (1) (716) 7Z'21066
Categories A1

Judith Eisen
P.O. Box 142
Guerneville, CA
95446
Telephone (1) (707) 869 3083
Categories A2

ERIM
P.O. Box 8618
Ann Arbor, MI 48107
Telephone (1) (313) 9941220
Contact: Ivan Cindrich
Categories Al

Bennett]. Feferman
3610 Misty Oak Dr ive, Apt. 140
Melbourne, FL 32901
Telephone (1) (407) 984 8894
Categories AI

Flatiron Studio
15 West 24th Street - 7th floo
New York, NY 10010
Telephone (1) (212) 6455173
Categories A2

Susannah D. Foltz
1824 Silver S.E.
Albuquerque, NM 87106
Telephone (1)(505) 2772616
Categories A2

Arthur David Fornari
813 Eighth Avenue
Brooklyn, NY 11215
Telephone (1) (718) 965 3956
Categories A2 A 7 A8 B5

Art Freund
124 Brookwood Drive
Santa Cruz, CA 95065
Telephone (1) (408) 4581619
Categories: A2 A3 A8

Global Images, Inc.
509 Madison Avenue, Suite 1400
NewYork, NY
lOOZ'2
Telephone (1) (212) 7598606
Contact: Walter Clarke
Categories A1

GPM
4165 Apalogen Road
Philadelphia, PA
19144
Telephone (1) (215) 8494049
Categories: A1 A4 A6 A7 A8 A9 A10 A11 A12 B7 8
B9

JoGoing
Mile 68 Richardson Highway
SR No. 10
Fairbanks, AK 99701
Categories AI

HOLOGRAM PRODUCERS

A1= Commercial H-1	**A5**=Holographic Stereograms	**A9**= Custom Lighting Support
A2=Fine Art Originals	**A6**=Embossed/Hotstamped	**A10**=Oversize/Large Format
A3=Fine Art Ltd Editions	**A7**=Silver Halide Transmission	**A11**=Pulse Holograms
A4=Custom Presentation Support	**A8**=Silver Halide Reflection	**A12**=Dichromate

Bruce Goldberg
509 Silver Ave.
San Francisco, CA
94112
Telephone (1) (415) 584 1192
Categories A2

Kenneth M. Goss
122 North Michigan
Belleville, IL 62221
Telephone (1) (618) 2774211
Categories A2

Matthew E. Hansen
741 East Gorham Street
Madison, WI
3703
Telephone (1)(608) 2553580
Categories A2

HTI - High Tech Imaging
1428 Darlington
Youngstown, OH
44505
USA
Telephone (1) (800) 458 6859
Categories: A2

Hoechst Celanese Corporation
86 Morris Avenue
Summit, NJ
07901
Telephone (1) (201)5227816
Contact:. Gunilla Gilberg
Categories: A1

Holage
1881 Eighth Avenue
San Francisco, CA
94122
Telephone (1) (415) 564 1840
Contact. Brad D. Cantos
Categories A3 A7 A8

Holaxis Corporation
968 Farmington Avenue
Hartford, CT
06107
Telephone (1) (203)232 2030
Contact: Martin Berson
Categories: A1 A4 A5 A6 A7 A8 A9 A10 A11 A12 B1 B3 B4 B5 B6 B7 B8 B9

Holicon Corporation
906 University Place
Evanston, IL
60201
Telephone (1) (312) 4914310
Contact: Dr. Hans Bjelkhagen
FAX (1) (312) 491 7955
Categories: A1 A2 A3 A7 A8 A9 A10 A11

Holicon Pulsed Laser Holography
P.O. Box 451
Lake Bluff, IL 60044
Telephone (1) (312) 2346633
Contact: Dr. Hans Bjelkhagen
FAX (1) (312) 491 7955
Categories: Al A2 A3 A7 A8 A9 A10 A11

Holocraft International
P.O. Box 152
Lake Forest, IL 60045
Telephone (1)(312) 2347625
Contact: William A. Christ
Categories A2

Holocrafts of Long Island
227 9th Street
West Babylon, NY 11704-3728
Telephone (1) (516) 6690372
Contact:. Pat Willard
Categories A2

HOLOGRAM PRODUCERS

A1= Commercial H-1
A2=Fine Art Originals
A3=Fine Art Ltd Editions
A4=Custom Presentation
　　Support

A5=Holographic Stereograms
A6=Embossed/Hotstamped
A7=Silver Halide Transmission
A8=Silver Halide Reflection

A9= Custom Lighting Support
A10=Oversize/Large Format
A11=Pulse Holograms
A12=Dichromate

Holoflex Company
RR3,Box381
Urbana, IL 61801
USA
Telephone (1)(217) 684 2102
Contact: Donald Barnhart
Categories: Al A5 A 7 A8

Holography Institute
P.O. Box 446
Petaluma, CA
94953
Telephone (1) (707) 7781497
Contact: P. Pink
Categories: A1 A3 A6 B7 C6 C7 C10 C12

Holografix
Attn: Sherwin Chew
431 Linda Avenue #lA
Piedmont, CA 94611
Telephone (1) (415) 6539925
Categories A1

Holograma
5832 Reemelin Road
Cincinnati,OH
45248-1631
USA
Telephone (1) (513) 5741423
Contact Steven Merritt
Categories A1

Holographic Applications
21 Woodland Way
Greenbelt, MD
20770
USA
Telephone (1) (301) 3454652
Contact: Suzanne St. Cyr
Categories: Al A2 A3 A4 A5 A6 A7
A8 A9 A10 A11 A12 C9 C8

Holographic Design, Inc.
1084 North Delaware Avenue
Philadelphia, PA
19125
USA
Telephone (1) (215) 4259220
Contact D.Miller
Categories: Al A4 A6 AI2 B3 B6 B7 B8 C7

Holographic Design Systems, Incorporated
1134 West Washington Blvd.
Chicago,IL
60607
Telephone (1) (312) 829 2292
Contact: Robert Billings
Categories: Al A2 A3 A4 AS A6 A7 A8 A9 A10 B7 B8

HOLOGRAM PRODUCERS

A1= Commercial H-1
A2=Fine Art Originals
A3=Fine Art Ltd Editions
A4=Custom Presentation
    Support

A5=Holographic Stereograms
A6=Embossed/Hotstamped
A7=Silver Halide Transmission
A8=Silver Halide Reflection

A9= Custom Lighting Support
A10=Oversize/Large Format
A11=Pulse Holograms
A12=Dichromate

Holographic Dimensions, Inc.
12247 S.w. 132 Court
Miami,FL
33186
Telephone (1) (305) 2554247
Contact:. Kevin G. Brown
Categories A1

Holographic Images, Inc.
Attn: Larry Lieberman
1301 Dade Blvd
Miami Beach, FL
33139
Telephone (1)(305) 5315465
Categories A1

Holographic Studios
240 East 26th Street
New York, NY 10010
Contact:. Jason
Telex:. (1)(212) 6869397
FAX: (1)(212) 4818645
Categories: A1 A2 A3 A4 AS A6 A7 A8 A9 A10 B1
B2 B3 B4 B5 B6 B7 B8 B9 C4 C10

Holographics Inc.
4401 llthSt
Long Island City, NY
11101
U&\
Telephone (1)(718) 7843435
Contact: Ana Maria Nicholson
Categories A1

Holographics North, Inc.
Attn: Dr. John Perry
444 South Union Street
Burlington, vr
05401
Telephone (1) (802) 568 2275
Categories: A2

Holography News
3932 Mckinley Street, N.W.
Washington, D.C.
20015
USA
Contact. L Kontnick
Categories A1 B7 C12

Holography Workshops
Lake Forest College
Sheridan and Collge Road
Lake Forest, IL
00045
Telephone (1) (312) 234 3100
Contact. Tung H.Jeong
Categories: A1 A2 A4 A7 A8 A9 A10 A11 A12 B1 C1
C2 C3 C4 C5 C6 C7 C8 C9 C10 C11 C12

Holomatrix Inc.
P.o. Box 808
Hillside, NJ
07205
USA
Telephone (1) (800)922 0958
Categories A1

Holo,spectra
7742-B Gloria Avenue
VanNuys,CA
91406
Telephone (1) (818) 9942577
Contact W Arkin
FAX: (1)(818) 9944709
Categories: A1 A4 A5 A6 A 7 A8 A9 A12 B3 B8 D2
D3 D4 D5 D6 D7 D8 D9 D10 D11 D12

Holowaves Hawaii
339 Saratoga Road
Honolulu, HI
96815
Contact: Cecilia Lemieux
FAX: (1)(808) 9241095
Categories A3 A8

HOLOGRAM PRODUCERS

A1= Commercial H-1
A2=Fine Art Originals
A3=Fine Art Ltd Editions
A4=Custom Presentation
   Support

A5=Holographic Stereograms
A6=Embossed/Hotstamped
A7=Silver Halide Transmission
A8=Silver Halide Reflection

A9= Custom Lighting Support
A10=Oversize/Large Format
A11=Pulse Holograms
A12=Dichromate

Hford Photo Corporation
Attn: Edward Sachtler
West 70 Century Road
Paramus, ~ 07652
Telephone (1) (201) 265 6000
Categories: A1

Images Company
P.O. Box 313
Jamaica, NY 11419
Telephone (1)(718) 706 5003
Categories A1

Imaging & Design
1101 Rcmsom Road
Grand Island, NY
14072-1459
Telephone (1) (716) 7737272
Contact. Keith Allen
Categories: A1 A4 A6 A7 A8 A9 A10 A11 B7 B8 B9

Institute of Optical Research
Attn: Matt Hannifin
2307Gulf
Midland, TX 79705
Telephone (1) (915) 6828740
Categories A2

Randy James/Holography
P.O. Box 305
Santa Cruz, CA
95061
Telephone (1) (408) 4584213
Contact. Randy James
Categories A1A7A8

John Kaufman Photography
P.O. Box 477
Point Reyes Station, CA 94956
Telephone (1) (415) 6631216
Contact John Kaufman
Categories: A2 A3 A8 AlO B5 C4

Aaron Kurzen
P.O. Box 3233
Stony Creek, cr 06405
Telephone (1) (203) 488 4711
Telex. Ph: (1) (212) 535 2180
Categories: A2

Lasart Inc
P.O. Box 4621
Laguna Beach, CA 92652
Telephone (1) (714) 499 5508
Contact August Muth
Categories: A2

LasartLtd.
P.O. Box 703 - RdAO.50
NOIWood, CO 81423
Telephone (1) (303) 3274701
Categories: A1

Laser Affiliates
2047 Blucher VaHey Road
Sebastopol, CA
95472
Telephone (1) (707) 8237171
Contact Nancy Gorglione
Categories: A2 A3 A4 A7 A8 A9 A10 C2 C3 C6 C9 C10

Laser Arts
1712 Catherdal
Plano, TX 75023
Telephone (1) (214) 4230158
Categories A1

LAS.E.R. Co.
1900 Gore Drive
Haymarket, VA 22069
Telephone (1) (703) 7542526
Contact Jim Bowman
Categories: A2 A3 A7 A8 A9 B2 B5 B9 C4 C12

HOLOGRAM PRODUCERS

A1= Commercial H-1	**A5**=Holographic Stereograms	**A9**= Custom Lighting Support
A2=Fine Art Originals	**A6**=Embossed/Hotstamped	**A10**=Oversize/Large Format
A3=Fine Art Ltd Editions	**A7**=Silver Halide Transmission	**A11**=Pulse Holograms
A4=Custom Presentation Support	**A8**=Silver Halide Reflection	**A12**=Dichromate

Laser Dreams
Attn: Nancy J. Gorglione
P.O. Box 326
Forestville, CA 95436
Telephone (1) (707) 8236104
Categories: A2 A3 A4 A7 A8 A9 AI0 B2 B3 B6 B9
C1

Laser Fare Ltd.
15 Industrial Lane
Johnston, RI 02919
Telephone (1)(401) 231 4400
Contact:. Terry Feely
Categories A2

Laser Ionics Inc.
701 South Kirkman Road
Orlando, FL
Telephone (1) (305) 298 1561
Categories A2

Lasermedia
2046 Armacost Ave.
Los Angeles, CA
90025
Telephone (1) (213) 820 3750
Categories: A1

The Lasersmiths
1000 West Monroe
Chicago,IL
60607
Telephone (1)(312) 7335462
Categories A1

Linda Law Holographies
8 Crescent Drive
Huntington, NY 11743
Telephone (1) (516) 351 6056
Contact:. linda Law
Categories: A2 A3 A7 A8 A9 C2 C3 C4 C6 C7 C8
C11 C12

Christopher J. LeSar
100 Hickory Lane
Lancaster, OH 43130
Telephone (1) (614) 6540862
Categories A2

Steven Lev
1871 Selma
Youngstown, OH 44503
Telephone (1) (216) 7475200
Categories: A2

Liconix
1390 Borregas Ave.
Sunnyvale, CA
94089
Telephone (1) (408) 734 4331
Categories: A2

Light Harmonics Inc.
93 Lake Shore Drive
Oakland, NJ
07436
Telephone (1)(201) 3378868
Contact: Jonathan Klempner
FAA': (1) (2014329542
Categories: A1 A2 A3 A5 A6 A7 A8 A9 AI0 All B2
B3 B5 B7 B9 C3 C4 C6 C7

Gail Light
1521 Revere Circle
Schaumburg, IL 60193
Telephone (1)(312) 3519545
Categories: A2

Light Impressions Inc.
149 B Josephine Street
Santa Cruz, CA 95060
Telephone (1) (408) 4581991
Contact: Kathryn S. Long, Mgr.
FAX: (1)(408) 4583338
Categories: A1 A5 A6 B2 B3 B4 B6 B7 C8

HOLOGRAM PRODUCERS

A1= Commercial H-1
A2=Fine Art Originals
A3=Fine Art Ltd Editions
A4=Custom Presentation
    Support

A5=Holographic Stereograms
A6=Embossed/Hotstamped
A7=Silver Halide Transmission
A8=Silver Halide Reflection

A9= Custom Lighting Support
A10=Oversize/Large Format
A11=Pulse Holograms
A12=Dichromate

Neal Lubetsky
654 Broadway
New York, NY 10012
Telephone (1) (212) 6748996
Categories A2A3

MacShane Holography
512 Braeside Drive
Arlington Heights, IL 60004
Telephone (1)(312) 3984983
Categories A2

Man Environment, Inc.
2041 Sawtelle Boulevard
Los Angeles, CA 90025
Telephone (1) (213) 4777922
Categories A2

Gerald Marks Studio
29 West 26th Street
New York, NY 10010
Telephone (1) (212) 8895994
Categories Al A2 A3 A4 AS C9

James A. McClean
21 Woodland Ave.
Butler, NJ 07405
Categories. A2

Don McNair
P.O. Box 173
Magnolia Springs, AL
36555
Telephone (1) (205) 965 7825
Categories: A2

Media Interface, Ltd.
167 Garfield Place
Brooklyn, NY 11215
Telephone (1) (718) 7884012
Contact: Ronald R Erickson
Categories: A1 A4 A9 B9 C3 C6

Melles Griot
1770 Kettering Street
Irvine, CA
92714
USA
Telephone (1)(714) 2615600
Categories A2

Michael G. Merrick
1002 Meadowview Drive
Mendota, IL
61342-1444
USA
Categories A2

Mark C. Merrill
410 Riverdale -Studio B
Glendale, CA
91204
USA
Telephone (1)(818) 2476458
Categories A2

Steve Moore
P.O. Box 148312
Chicago,IL
60657
USA
Telephone (1)(312) 871 6469
Categories A2

Aaron Muskovitz
P.O. Box 1022
South Lake Tahoe, CA 95705
Telephone (1)(916) 544 5989
Categories A2

Michael Naimark
216 Filbert Street
San Francisco, CA 94133
Telephone (1)(415) 3914817
Categories A2

HOLOGRAM PRODUCERS

A1= Commercial H-1	**A5**=Holographic Stereograms	**A9**= Custom Lighting Support
A2=Fine Art Originals	**A6**=Embossed/Hotstamped	**A10**=Oversize/Large Format
A3=Fine Art Ltd Editions	**A7**=Silver Halide Transmission	**A11**=Pulse Holograms
A4=Custom Presentation Support	**A8**=Silver Halide Reflection	**A12**=Dichromate

Neovision Productions
P.O. Box 74227
Los Angeles, CA
90004
USA
Telephone (1)(213) 3870461
Categories: A2

New York Holographic Lab
P.O.Box 20391
Tomkins Square Station
New York, NY
10009
USA
Telephone (1) (212) 2549774
Contact:. Dan &hweitzer
FAX: (1) (212) 6741007
Categories: Al A2 A3 A6 A7 A8 B1 B5 C3 C4

North American Holographics
111 East &ranton Ave
Lake Bluff, IL
60044
USA
Telephone (1)(312) 3244244
Categories Al

North East Laboratories
1079 Mechanics Valley Road #35
North East, MD
21901
USA
Telephone (1) (301) 3982295
FAX: (1) (301) 3982295
Categories: A1, A6, A8

Odhner Holographics
833 Laurel Avenue
Orlando, FL
32803
Telephone (1) (407) 894 7966
Contact Jefferson Odhner
Categories: Al A2 A7 A8 C3 C4 C6 C8 C9 C12

Pacific Holographics, Inc.
Attn: Craig Klein
P.O. Box 39542
344 7 Laclede Avneue
Los Angeles, CA
90039
USA
Telephone (1) (213) 666 7680
Categories: A2

W.R Padnos
2019 North Damen Avenue
Chicago, IL
60647
USA
Telephone (1) (312) 3842647
Contact:. WR Padnos
Categories: A2 A4 A7 A8 A9

Joel Petersen
7343 Adams Street
Paramount, CA
90723
USA
Telephone (1) (213) 634 0434
Categories: A2

Antonio Peticov
712 Broadway
New York, NY 10003
Telephone (1)(212) 529 0465
Categories A2

Point of View Dimensions
654 Broadway
New York, NY
10012
USA
Telephone (1) (212) 6748996
Categories: Al AS A6 A7 A8 A10

HOLOGRAM PRODUCERS

A1= Commercial H-1
A2=Fine Art Originals
A3=Fine Art Ltd Editions
A4=Custom Presentation
    Support

A5=Holographic Stereograms
A6=Embossed/Hotstamped
A7=Silver Halide Transmission
A8=Silver Halide Reflection

A9= Custom Lighting Support
A10=Oversize/Large Format
A11=Pulse Holograms
A12=Dichromate

Polaroid Corporation
Research Laboratories
750 Main Street # lA
Cambridge, MA
02139
USA
Telephone (1)(617) 5773509
Contact: J. Cowan
Telex Tel: (I) (617) 5774169
FAX: Tel: (I) (617) 5773805
Categories: A12 Dl D6

Portson Inc.
9201 Quivira
Overland Park, KS
66215-3905
USA
Telephone (1) (913) 4927010
Contact:. Jill Jarvis
FAX: (1)(913) 4927099
Categories: Al A4 A7 A8 AI0 A12 B3

Mr. Richard Rallison
Ralcon
Box 142
8501 South 400 West
Paradise, UT 84328
Telephone (1) (801) 2454623
Categories A2

Reconnaissance
3932 McKinley Street, N.W.
Washington, D.C. 20015
Contact: L. Kontnick
FAX: (1)(703) 764-6398
Categories A1 B7 C12

Regal Press Inc.
129 Guild St
Norwood, MA
02062
Telephone (1)(617) 769 3900
Categories: Al

Alan Rhody
1639 Fulton Street Apt B
San Francisco, CA 94117
Telephone (1)(415) 567 8692
Categories: Al A2 A3 A4 A 7 A8 A9

Adel Rootstein, Inc.
205 West 19th Street
New York, NY
10017
Categories: A2

Michael P. Rosewell
2302 South Damen Avenue
Chicago,IL
60608
USA
Telephone (1) (312) 2542577
Categories: A2

E.C. Schultz & Co.
333 Crossen
Elk Grove Village, IL
60007
USA
Telephone (1) (312) 6401190
Contact: Lynn Schultz
Categories A3

Robert Sherwood Holographic Design
400 West Erie Street
Chicago, IL
60610
USA
Telephone (1)(312) 944 0784
Contact: K.Kellison,R.Sherwood,
FAX: (1) (215) 4259221
Categories: Al A4 A6 A12 B6 B7 B8 B9 C7 C11

HOLOGRAM PRODUCERS

A1= Commercial H-1
A2=Fine Art Originals
A3=Fine Art Ltd Editions
A4=Custom Presentation
         Support

A5=Holographic Stereograms
A6=Embossed/Hotstamped
A7=Silver Halide Transmission
A8=Silver Halide Reflection

A9= Custom Lighting Support
A10=Oversize/Large Format
A11=Pulse Holograms
A12=Dichromate

Catherine J. Smith
805 South Oakley Boulevard
Chicago, IL
60612
USA
Telephone (1)(312) 421 7225
Categories: A2

Jascha Sonis
227 Summit Ave. E-302
Brookline, MA
02146
Telephone (1) (802)496 2576
Categories A1

Space Age Designs
P.O. Box 72
Carversville, PA
18913
USA
Telephone (1) (215) 2978490
Contact Valli Rothaus
Categories A2 A3 A12 B3 B7

Spectratek Corp
Attn: Mike Wanlass
P.O. Box 3407
Culver City, CA
90230
USA
Telephone (1) (213) 836 4343
Categories A1

Steve Provence Holography
Represented by Holo/Source
21800 Melrose Avenue-Suite 7
Southfield, MI
48075
USA
Telephone (1) (313) 3550412
Contact Rob Levy
FAX: (1)(408) 3389833
Categories Al AS A6

Len Stockier
7227 Eastwood Street
Philadelphia, PA
19149
USA
Telephone (1) (215) 3315067
Categories A2

Synchronicity Holograms
Box 4235
Lincolnville, ME
04849
USA
Telephone (1) (207) 7633182
Contact Arlene Jurewicz
FAX: (1)(207) 2363847
Categories: A8 B4 C4 C6 C7 C8 C9 C10 Cll C12

Third Dimension Arts, LTD
3-D Arts T .M.
1241 Andersen Drive
Suites C&D
San Raphael, CA
94901
USA
Telephone (1)(415) 485 1730
Contact Tim Laduca
FAX: (1)(415) 4850435
Categories A6 A12 B2 B3 B4 B8

Donald K. Thornton
1175 Danielson Avenue
North Scituate, RI
02857-9219
USA
Telephone (1) (401) 6472608
Categories A2

KhiemTran
2126 West Haddon #2F
Chicago, IL 60622
Telephone (1) (312) 7726999
Categories A2

HOLOGRAM PRODUCERS

A1= Commercial H-1
A2=Fine Art Originals
A3=Fine Art Ltd Editions
A4=Custom Presentation
Support

A5=Holographic Stereograms
A6=Embossed/Hotstamped
A7=Silver Halide Transmission
A8=Silver Halide Reflection

A9= Custom Lighting Support
A10=Oversize/Large Format
A11=Pulse Holograms
A12=Dichromate

Transfer Print Foils, Inc.
9 Cotters Lane
P.O. Box 518
East Brunswick, NJ
08816
USA
Telephone (1) (201)2381800
Categories A6

Two Guys Holography
P.O. Box 148312
Chicago,IL
60657
USA
Telephone (1) (312) 871 6468
Categories A2

Douglas E. Tyler
III North Second Street
Niles, MI
49120
USA
Telephone (1) (616) 6830934
Categories A2

Bill Vance
1277 Roxbury Road
Rockford, IL
61107
USA
Telephone (1)(815) 2821141
Categories A2

Heidi S. von der Gathen
5941 North Sacramento
Chicago,IL
60659
USA
Telephone (1) (312) 271 2176
Categories: A2

Eric Van Hamersveld
22030 Grant Avenue
Torrance, CA
90503
Telephone (1) (213) 9567469
Categories A2

Doris Vila
157 East 33rd Street
New York, NY
10016
USA
Telephone (1) (212) 686 5387
Categories: Al A2 A3 A4 A7 A9 A10 C4 C5 C6 C9

Wave Guides Inc.
521 East 12th Street #11
New York, NY
10009
Telephone (1) (212) 9823713
Contact Hale Aust
Categories: Al A2 A3 A4 A5 A 7 A8 A9 A10

Sally Weber
13230 Leach Street
Sylmar, CA
91342
Telephone (1) (818) 3624637
Categories A2 A3 A7 A10

Clifford B Weissman
609 Greenlawn Avenue
Dayton, OH
45403
Telephone (1) (513) 2552809
Categories A2

Sandrajean Wells
P.O. Box 927
Federal Station
Worcester, MA
01601
Categories A2

HOLOGRAM PRODUCERS

A1= Commercial H-1
A2=Fine Art Originals
A3=Fine Art Ltd Editions
A4=Custom Presentation
        Support

A5=Holographic Stereograms
A6=Embossed/Hotstamped
A7=Silver Halide Transmission
A8=Silver Halide Reflection

A9= Custom Lighting Support
A10=Oversize/Large Format
A11=Pulse Holograms
A12=Dichromate

Ed Wesly
5331 North Kenmore Avenue
Chicago, IL
60640
USA
Telephone (I) (312) 7841669
Categories: A2 A3 A7 A8 A11 C1 C3 C4 C5 C6 C7
C8 C9 C10 C11 C12

White Light Works, Inc.
P.O. Box 851
Woodland Hills, CA
91365
USA
Telephone (1)(818) 7031111
Contact: J erry Fox
FAX: (1)(818) 7031182
Categories: A1 A4 AS A6 A7 A8 A12

WholeHogGraphy
4142 BellefontA1ne
Houston, TX
77025-1105
USA
Telephone (1) (713) 6673325
Contact M. Crawford
Categories: A2 A3 AS A 7 A8

Whole Message Studios
919 West Mission Street
Santa Barbara, CA
93101
USA
Telephone (1) (805) 6822345
Categories A2

Carol Yeager
RD 2 Box 518
Catskills, NY 12414
Telephone (1) (518) 9432007
Categories A1

Lee Zeman
55 Ann Street
New York, NY 10038
Telephone (I) (212) 7321854
Categories: A1

WEST GERMANY
Detlev Abendroth
PotsdamerStrasselO
4300 Essen 1
Federal Republic of Germany
Telephone (49)(201) 742233
Categories A1

Arbeitskreis Holografie B.V.
Herman;J osef Bianchi
Boeckelter WEG 47
4170 Geldem
Federal Republic of Germany
Telephone (49) (2831) 3034
Categories A1

Atomika Technische Physik GmbH
0..8000 Munchen 19
Kuglmuellerstrasse 6
Federal Republic of Germany
Telephone (49) (89) 152031
Categories A1

Carmenza Dominguez
Schweizerstrasse 77
6000 Frankfurt 70
Federal Republic of Germany
Telephone (49) (69) 627836
Categories A1

HOLOGRAM PRODUCERS

A1= Commercial H-1
A2=Fine Art Originals
A3=Fine Art Ltd Editions
A4=Custom Presentation Support

A5=Holographic Stereograms
A6=Embossed/Hotstamped
A7=Silver Halide Transmission
A8=Silver Halide Reflection

A9= Custom Lighting Support
A10=Oversize/Large Format
A11=Pulse Holograms
A12=Dichromate

Julian Fischer
Klugstrasse 49
8000 Munchen 19, West Germany
Telephone (49) (83) 1572682
Categories A1

Martin Hofmann
Carl-hermann-Gaiserstrasse 20
7320 Groppingen
Federal Republic of Germany
Telephone (49)(7161) 12200
Categories A1

Holar Seele KG
Wasserwerksweg 10-14
0.2960 Aurich 1
Federal Republic ofGerrnany
Telephone (49) (41) 10005
Categories A1

Holografie - Hofmann Labor
C.H. Gdiserstrasse 20
0.7320 Goppingen
Federal Republic of Germany
Contact: Mr. Hofmann
Telephone (49) (07161) 12200
Categories: A1 A2 A3 A4 A5 A6 A 7 A8 A9 A10 A11
A12 B1 B2 B3 B4 B5 B6 B7 B8 B9 C4 C6 D1

HOLO GmbH
Holografielabor Osnabruck
MindemerStr. 205
0.4500 Osnabruck
Federal Republic of Germany
Telephone (0) (541) 7102173
Contact: Vito Orazem, T .Luck
FAX: (0)(541) 7102176
Categories: Al A2 A3 A4 A 7 A8 A9 A10 B2 B3 B4
B5 B7 B9 C3 C4 C7 C10

Holographie Labor
Georgenstrasse 61
D-8000 Munich 40
Federal Republic of Germany
Telephone (49) (89) 271 2989
Telephone (49) (89) 4481707
Contact. M. Mielke
FAX: (49)(89) 2711375
Categories: Al A2 A3 A5 A7 A8 A10 A12 B2 B5

Dieter Jung
Viuonvillestrasse 11
D 10000 Berlin 41
Federal Republic of Germany
Telephone (49) (30) 7718431
Categories A1

Andreas Kaufman
Graf-von-Galen Strasse 5
0.4800 Bielefeld 1,
Federal Republic of Germany
Telephone (49) (521) 102269
Categories A1

Labor fur Holografie
AmmForst38
0.4230 Wesel
Federal Republic of Germany
Telephone (49) (281) 52837
Categories A1

Laserion Handels GmbH
Postfach 110268
2800 Bremen 11
Federal Republic of Germany
Telephone (49) (421) 449550
Categories A1

HOLOGRAM PRODUCERS

A1= Commercial H-1
A2=Fine Art Originals
A3=Fine Art Ltd Editions
A4=Custom Presentation
        Support

A5=Holographic Stereograms
A6=Embossed/Hotstamped
A7=Silver Halide Transmission
A8=Silver Halide Reflection

A9= Custom Lighting Support
A10=Oversize/Large Format
A11=Pulse Holograms
A12=Dichromate

UdoLaube
Robert Koch Strasse 41
6100 Darmstadt,
Federal Republic of Germany
Telephone (49) (6151) 537181
Categories: A1

Merlin
AchimKonz
Wassenberger Strasse 47
5138 Heinsberg
Federal Republic of Germany
Telephone (49) (2452) 6072
Categories Al

Moeller Wedel Optische Werk
Rosengarten 10
2000 Wedel
Federal Republic of Germany
Categories: Al

Spindeler & Hoyer GmbH
Koenigsallee 23
0-3400 Goettingen
Federal Republic of Germany
Categories A1

Stein bichler Labor
Am Bauhof4
0-8201 Neubeuem
Federal Republic of Germany
Telephone (49)(8035) 1017.
Categories: Al

Mirau Volker
Schosserstrasse 93
46 Dortmund I
Federal Republic of Germany
Telephone (49) (4) 186 7413
Categories A1

Late entry:
Holo Media AB/Hologram Museum
Box 45012
Drottninggatan 100
10430 Stockholm
Sweden
Telephone (46) (08) 105465
Contact: Mona Forsberg
FAX: (46) (08) 107638
Categories: A1, A6, A11, A12, B1, B2, B3, B4, B5, B6,
B7, B8, B9, C1, C8, C10

CHAPTER 5

BUYING AND SELLING

This chapter concerns holograms already produced for sale. This is an overview of the wide variety of holograms available for sale on the commercial level and the limited edition (one-of-a-kind) fine art holograms which are usually in the collector's market. This section also discusses areas in which the medium can improve and expand. The glossary and chapter one give a basic presentation of different types of holograms.

From the chart and names and addresses sections following this chapter you will be able to find the many and growing list of people from whom holograms are available for sale. The chart lists all the companies and checks which buying and selling categories they fit into. In the names and addresses section, all categories that a company checks off are listed- in the Producer, Buying & Selling, Education, and Equipment & Supplies areas. Branch offices have separate listings, so there may appear to be duplicates, but their addresses are in fact different. Companies are listed alphabetically within their country.

As with many industries, you will find that merchandise in this trade moves from Producer (company or individual) to large distributor to retail shop to customer, or by direct mail to the customer. In general, it works like this: the producer decides the retail price of the hologram. Retail shops buy holograms at 50% off the retail price. Distributors want to make about 30% on whatever they buy from the producer and sell to the retail shop.

Producer

If a producer shows a hologram to a distributor and the distributor wants to sell the product, one of the first topics for discussion is exclusive rights. The distributor generally asks for exclusive territorial or global rights. Everything is negotiable in this industry and there are no firm rules.

There are advantages and disadvantages to exclusive rights. One advantage is that someone else does all the advertising and selling of your product. A disadvantage is that you give up control of your market.

If you are a producer and you want commissioned sales people, the best area to find them in this trade is through the gift industry. There are gift conventions in every major city attended by commissioned sales people and sales groups representing several companies.

Distributor

Distributors are often producers, retailers and direct mail marketers as well. They generally have their own (free) catalogue and separate wholesale and retail price lists. Items are generally sold "buyer beware" and are not returnable. However, a good, established retail customer with high volume orders can generally work out a more flexible arrangement.

Retailer

When retailers purchase merchandise from distributors or producers, the items are sold with a "suggested retail price" and generally arrive with the price unmarked. The individual retail shops then establish their own selling price. The number of retail shops that carry a good selection of holograms has grown dramatically in the last few years. Many of the most successful shops, as you might expect, are in high traffic areas. Holograms are welcomed as a sideline by many shops because a well lit display of holograms has proven one of the best traffic stoppers around.

What Holograms for what price?

No matter how well you may think holography would work in a certain area, the bottom line always seems to be the price matching the application. Embossed holograms have finally become inexpensive enough to produce a "give away" item and have been used in this way by some companies.

HOLOGRAMS FOR UNDER (US) $10

Holograms suggested to sell from $1 to $10 include not only embossed holograms but also silver halide reflection holograms and some dichromate reflection holograms. Many holograms at this level of wholesale price are pre-packaged for sale as a framed hologram, a piece of jewelry with its own display or packaged in material such as a paper weight.

Embossed holograms under $10

This area of the stock hologram market has undergone major improvements. For those who would like a very good mass-manufactured hologram, these embossed holograms are presented reasonably well for a price to fit just about everybody's budget. In this price range the embossed holograms are sometimes larger (up to six square inches), usually enclosed in cellophane with information briefly explaining the hologram. Although the embossing technology can produce high quality, the images have. improved only slightly and are not very Impressive on the whole. For those to whom holography is still new the embossed hologram is a nice, inexpensive take home item and is a good example of mirror-backed transmission holography. In summary, the technology of custom-produced images has improved tremendously in the embossed hologram industry but the stock images have been slow in upgrading.

Silver halide reflection holograms under $10

Silver halide reflection holograms have moved down to an affordable price range as everyone knew they could. They have also, for the most part, not compromised much on image quality. These holograms are usually slightly over two inches square. They are often matted and neatly framed in an inexpensive plastic frame that is ready to hang or to be placed on a horizontal surface. The technical quality can be described as very bright, sharp and with clever images. The images are exemplary of the different techniques of reflection holography like color shifting, animated multi-channel holograms, simple multi-channel image changes and effective image-planed images that are as sharp in front as they are behind the hologram surface. These holograms are on triacetate or polyester base silver halide film and are well worth a buyer's further investigation.

Dichromate reflection holograms under $10

The dichromate reflection holograms that have entered the under-ten-dollar retail bracket are sometimes bright, sharp and even interesting. Sometimes they are not. At these prices the subject matter is limited but hopefully this will change soon. To be quite honest, the holograms at the next price level up are not very lively and varied in subject matter either, so the lower price is at least some good news.

Dichromate holograms for under $10 are showing up on items like jewelry pendants and key chains and are usually small circles, an inch and a half in diameter with clear backing. It is sometimes unfortunate that this is where many people get their initial impressions of the dichromate hologram. However if this very low price for dichromate holography can bring up the quality, it deserves to disrupt the competition.

HOLOGRAMS FROM (US) $10- $30

The selection in the ten to thirty dollar range has also increased in every way, including overall quality. Embossed holograms are usually out of this price range unless it is a special item because of size, limited edition or unique application. This price range belongs more to silver halide and dichromate reflection holograms.

Silver halide reflection holograms from $10 - $30

The silver halide reflection holograms have a wider selection of images in this retail price range with sizes over the standard four by five inch format yet under the eight by ten inch format.

The detail and intricacy of subject selection is better

explored in a larger size and many holographic techniques are also manifest in this pricing area. The framing is more substantial with metal and glass replacing plastic frames. These images would be significantly cheaper without the pre-framing but having the hologram ready for the wall does save time and aggravation if you are not prepared to frame it yourself. These holograms are mass-manufactured on film and consist of what are hoped to be popular images.

Dichromate reflection holograms from $10 - $30
The dichromate reflection holograms in this price range are typical images set into paper weights, key chains, pendants and novelty items (like those psycho-eyeball hologram-lensed glasses). Watches whose crystal covers have a hologram are also a big seller in this category. Non- pendant holograms are as large as two by three inches in a thin glass sandwich. The dichromate reflection hologram is an excellent hologram for so many subjects that are of shallow depth. It can hold tremendous detail under the widest range of available light. When you see one you will know.

HOLOGRAMS FROM (US) $30 - $50
The thirty to fifty dollar price range includes basically the same items as in the previous category. The difference exists in more complicated techniques, slightly larger image and size and the rare inclusion of glass plates as the hologram recording material. Some excellent holographic jewelry with more original subject handling begins in this category. Recently, more of the cottage industry has been moving into this category. This is also the range for lower priced artists' holographic images. They most likely will not be limited edition holograms, but rather popular images dressed up for a better sale. For example, sometimes the frames are of a custom shape and hand-painted to suit the image.

HOLOGRAMS FROM (US) $50 - $100
At the fifty to one hundred dollar range the major hologram types are still silver halide reflection and dichromate reflection with the possibility of some artists' limited edition work. The hologram sizes for these stock holograms have risen to as much as
eight by ten inches and more specialty items can be included .

Less than five years ago some examples of silver halide transmission holograms would also have been included within this price range. Reflection holog-

raphy has nearly eliminated this type of hologram from the stock hologram market. Reflection holograms hang on the wall like pictures while transmission holograms, even if white-light viewable, are more of a nuisance to light. This is very unfortunate because white light transmission holograms have much greater depth potentials and often have more flexible image combinations than reflection holograms.

There are self-contained lighting units for the white light transmission stereogram. In the same way, there could also be self-contained lighting units designed for any type of white light transmission holograms at a cheaper cost today. We note that some reflection holograms in this price range have complementary lighting/display units built with the hologram and are sold ready to view.

HOLOGRAMS FOR OVER (US) $100
Past the hundred dollar limit you can expect stock images to include: the more exotic pulse hologram subjects, all major examples of the holographic stereograms and all hologram sizes up to the largest standard formats in holographic recording materials. These stock image prices vary widely and there are no standard price ranges really worth mentioning. As in any multiple medium, much depends upon the studio, the artist/designer and the subject almost as much as on the quality of the individual, technical work.

FINE ART HOLOGRAMS & COLLECTIBLES
Once in this price range it is worth considering any of the many limited edition artist works that become available. From an historical point of view alone, these early holograms by artists could in time be worth incredible sums of money if our current collectors' market is any indicator. Regardless of art or aesthetic content, the fact that these are early holograms is worth an investment.

It has been determined by financial analysts that art was one of the best markets to have invested in during the last decade. Art is one of few investments that seems to have weathered the stock market crash well. Also to the buyers' advantage is that while paintings and sculptures by living artists can sell for literally hundreds of thousands of dollars, you can collect the absolute best holograms for comparatively very little.

General Advice

TIPS ON AUTHENTICITY OF HOLOGRAMS
Be sure to inspect the art holograms you purchase

for the signature of the artist and the edition numbers. Be sure that you know ahead of time if it is a limited edition, how many have been made in the edition and get written assurances that this hologram will only be reproduced up to that edition number. Also ask if the hologram has been sealed and by what means. Some commercial houses use a material that looks like contact paper and is not a reliable, long-term protectant. If properly taken care of, holograms should last longer than the photographs from the nineteenth century.

PROPER HOLOGRAM CARE

Proper care of holograms is largely common sense. Do not put holograms in direct sunlight or any other source of ultraviolet light for a prolonged time. This light destroys recording material faster than normal lighting conditions. Heat and humidity should also be avoided. All holograms should have their emulsions sealed off to avoid contamination by air and moisture especially in warmer climates. Film holograms of triacetate or polyester should be cleaned, if necessary, by professionals familiar with conservation of photosensitive materials and holography. Because someone knows about photography is no assurance they know the correct way to deal with holography. Holographic processing yeilds an entirely different result than photography.

PRESENT HOLOGRAMS WELL

In both the commercial stock image area of holography and in the fine art area, the development of the medium has been slowed by two closely related areas: proper marketing and education. Every medium has its parameters. Holography is very different from most other recording or image-making media by the way it is made, the way it is best illuminated and the fact that it is capable of displaying in the most minute detail the reflected light of any object.

When a hologram is executed well it is very sad to see it displayed in poor lighting or none at all. Some holograms placed on objects like paperweights or toys are vary good in concept and application but will not live up to reasonable expectations if retailers do not display or illuminate them correctly.

MARKET HOLOGRAMS WITH PRODUCT KNOWLEDGE

Education is another area that needs improvement. Marketing people need to know what they are in fact marketing to make the right decisions. More

people need to know what a hologram is and what it can be in order to become excited about it. Even the pet rock had a bit of background information about it .

The stock hologram sellers are getting the message and are offering inexpensive displays complete with simple but effective lighting for their lines of images. Some packaging now includes information about holography and how to display it. There is still some more work to be done in this area before holography truly takes hold. More and better displays coupled with a more serious effort to educate the people buying and selling holograms will take holography over its last major hurdles. When anyone sees this medium done well it is very hard not to like it.

HOLOGRAM PRESERVATION

Silver halide, Dichromate and PhotapoEymer holograms

The jury is still out on the one perfect way to present and protect holograms made in the silver halide gelatin emulsions so no recommendations will be made here. All holograms should be sealed from the atmosphere and humidity. Try to avoid putting the actual emulsion of the finished hologram in contact with any materials like hardware store variety spray paint, wood, wood materials, waterbase fluids, paper (even acid-free paper holds moisture and has fibers), rubber, tape, glass (moisture condenses on glass) and nearly all adhesives. Most holographic emulsions are destroyed when in prolonged contact with water vapor.

Once sealed properly, keep the holograms away from long exposure to direct sunlight (ultraviolet rays are very harmful over time) and avoid very hot and humid climate conditions. If you use wood frames or a mat, it is a good idea to seal the inside to keep the acids away from the hologram seal.

Some producers use a special anti-ultraviolet light, and coated glass on the face of the hologram on exhibition. This is expensive and may not be altogether practical. Using light sources with no UV light is just as effective. Clear plastic or acrylic on the front of the hologram is not advised, as these materials are easily scratched and attract dirt and dust. If you plan to ship a museum-quality hologram, often a hard plastic face cover is recommended over glass because of its resistance to breakage. Wood frames also take shock from the perimeter much better than metal frames in shipping. If you have a reflection hologram that is to be framed, back it with foamcore board and seal the back of the frame with archival tape.

Archival presentation materials can be found in several mail-order locations and in the better art supply stores. Call up the nearest museum for suggestions of where to find the one nearest you. The conservation department is the best place to inquire.

If you are a commercial user of holograms and the longest life of the hologram is not all that important there are several reasonable shortcuts to take. Your holographer or distributor can offer some good suggestions. Even if you have an inexpensive hologram that is unsigned you may still want it to last for generations. Follow the most obvious advice here and you should eliminate most problems. Obviously, there are some more exotic solutions for preserving very special and expensive pieces.

Embossed holograms

Embossed holography should last longer than this earth in most cases. It just needs to be prevented from surface scratches. You will notice that your credit card hologram wears quite well under circumstances that would probably destroy other hologram types. The only thing to really watch out for other than scratches is putting this hologram onto a surface that is textured. The embossed hologram, especially if its embossing material is thin, can take on the texture of the surface to which it is adhered, and thus destroy the image. The ripple finish prevents the image from displaying. The embossed hologram is extremely versatile, easy to take care of and very easy to present on any flat surface.

HOLOGRAM DISPLAY

Shop Displays

Before making a commercial display for your hologram consider your audience. Do you want shorter, child-size viewing heights or do you want taller, adult-size heights? Often a child will not be able to see the best lit hologram because it is placed at the wrong height. The best solution, if possible, is to place holograms at varying heights which makes them viewable by all customers.

Holograms attract attention when displayed correctly in high-traffic areas. At worst, shop owners would be sued for whiplash as people of all ages stop dead in their tracks doing a double or triple-take at an impressive holographic display. Educate the salespeople displaying the holographic products on how to maximize on holography's effectiveness as an attention-

getter. Again, many companies are now providing inexpensive hologram display units for their products that display the holograms well. If they have such displays, take advantage of them. They work well and save you time.

Despite having these displays there are still some people, even in science museums, who put extra lighting onto the holograms thinking the holograms will look better. They do not. Use the display in an area that will not be affected by other direct light sources. If you need a display unit for a Cross hologram, go to the hologram producer and you will find exactly what you need in any curved size. As with any other question in holography, the easiest solution is to call a consultant or the producer and ask. Often he or she can direct you to the best local sources with the best price for what you need.

Fine Art Displays

Fine Art holograms are more flexible in where they can be displayed because you are dealing with each one individually and usually have a more consistent height of viewing audience.

Reflection holograms are easily placed upon the wall like anyone-sided artwork. The non-mirrorbacked transmission hologram has to be displayed so light enters from behind at the correct angle. This means usually that the hologram is hanging into the room. Artists typically do not put frames on these holograms because they prefer them "floating". Also, most frames are not finished nicely in the back and it is easy to look through the hologram and see how the clips were attached. This type of frame is especially unsuitable for a transmission hologram.

The sealed hologram can be placed in a clear sandwich frame with holes at the top for ceiling hanging or placed into a slot on a pedestal or other bottom-support. Some also have displayed transmission holograms on panels, into which a custom-size hole is cut to fit the hologram. This allows the light to come from behind and offers support on all sides. People do bump into the freestanding edges of holograms in an exhibition if the space where the hologram is hanging has not been protected against the traffic. There are many curators of holography exhibitions listed in the Education names and addresses section who would be very helpful on important display matters for Fine Art holograms.

ILLUMINATION

Correct illumination of holograms is imperative but

it does not have to be a frighteningly expensive or confusing ordeal. If the hologram you have is white light-viewable this is especially true.

The correct illumination of holograms is critical to the success of the reconstruction of the image. Correct lighting not only depends on using the right illumination source but also on the angle and distance of illumination. Most holograms created in the latter half of this decade have a standard illumination angle of forty-five degrees. This angle usually is from above the hologram. If you are using a single direct source of light for the hologram at the forty-five degree angle from above that covers all of the hologram with strong, even light, the hologram should look great.

Laser-viewable holograms

If you have a laser-viewable hologram that you wish to display, you might have to hire a laser safety operator to do the lighting, register the laser properly and have the display certified by the appropriate authorities. The other option is to use a highly filtered light source like the sodium vapor or mercury arc lamps. These lighting units are expensive, need to be set up by a professional, are dangerous to the uninitiated and have a very powerful energy source. Unless the hologram is very large and worth the effort, get a white-light transfer made. Again, consult the holographer about this type of lighting as it is not the typical route.

Find out from the holographer the correct light source to be used. It can make a significant difference. Better lighting stores will have the MR-16-type, screw-in halogen point source lamps with varying light spreads. The bulbs sell for under twenty dollars and various fixtures are available to accommodate the bulb. This is an excellent lamp for reflection holograms.

White light-viewable holograms

For white light-viewable transmission holograms, a clear incandescent bulb with a single, horizontal filament is a good inexpensive light source. This type of bulb is available at almost any hardware store. If you use a hooded clamp light, the inside of the hood can be sprayed with flat black paint to get more direct light. By covering the hood with a piece of thin, blackened metal that has a rectangular opening cut out you can shape the light to cover the perimeter of the hologram inexpensively and effectively.

With a bigger budget, there are more handsome lighting designs whose units have built-in windows to frame the light and lens-controlled lighting. These lights come in a variety of designs for the way the light functions and in the way that the light is attached. Focused lighting is a beautiful way to illuminate holograms, especially if the light assemblage has the framing ability. These lights are typically over a hundred dollars apiece.

Holograms do not have to be displayed in totally dark rooms with black walls. Indirect atmosphere lighting is much nicer and the images still display very well. The transmission holograms look stronger when the background light does not compete with them. Reflection holograms with blackened backs can be hung on any color or texture of wall and look very good when they are lit well.

Holographic stereograms

White light-viewable Cross holographic stereograms usually have their own display and light bulb included. These holograms have a clear, incandescent bulb with a vertical filament to match the the composite of vertical slits. The Holodisk® is illuminated from directly above with a point-source light. The horizontal and the vertical filament clear bulbs cost less than five dollars each and are available at most stores.

Embossed, Dichromate, and Photopolymer holograms

Good embossed and dichromate holograms that are not too deep should show up in most lighting conditions. The source of light does not have to be too strong with these holograms and they are often too bright to look at if the source is too strong. What hurts the viewing of these holograms is multiple light sources and placement in dim, indirect light. Any clear, incandescent point-source bulb placed at the correct forty-five-degree angle will make these holograms display well.

LIGHT FANTASTIC

PLC

EST.1981

THE WHOLE PICTURE...

◯ LEISURE/TOURIST ATTRACTIONS

PICCADILLY CIRCUS – LONDON	THEME PARK NICE – FRANCE
COVENT GARDEN – LONDON	ARNDALE CENTRE – MANCHESTER
FUTUREWORLD STERLING – SCOTLAND	
NORTHERN LIGHT FANTASTIC EDINBURGH – SCOTLAND	
FRANKFURT / MUNICH – WEST GERMANY	
ONTARIO SCIENCE CENTRE – CANADA	

◯ RETAIL

AT ABOVE LOCATIONS

◯ WHOLESALE/DISTRIBUTION

TAIWAN	FRANCE	SPAIN
JAPAN	HOLLAND	GREECE
AUSTRALIA	ITALY	ISRAEL
U.S.A.	FINLAND	TURKEY
MEXICO	BELGIUM	INDIA
CANADA	SWEDEN	SAUDI ARABIA
BRAZIL	AUSTRIA	NORWAY
W. GERMANY	SWITZERLAND	KOREA

◯ COMMERCIAL (EMBOSSED HOLOGRAMS)

A SMALL SELECTION OF OUR CUSTOMERS WORLDWIDE

CAMEL	TYROLIA	ESTEE LAUDER
CITIBANK	LEGO	ROLLS ROYCE
PHILIP MORRIS	BRITISH TELECOM	B.B.C.
PERRIER	ROLEX	UVEX
AUSTRIAN STAMP CO.	LEONHARD KURZ	B.P.

...THE TOTAL SERVICE

LEISURE RETAIL DIVISION
TERRY A. WOODD
LIGHT FANTASTIC PLC, 48 SOUTH ROW,
THE MARKET, COVENT GARDEN,
LONDON WC2E 8HN, ENGLAND
TELEPHONE: 1-836 6423   FAX: 1-836 6429

COMMERCIAL DIVISION
ROGER C. KNIGHT
LIGHT FANTASTIC PLC,
GELDERS HALL ROAD, SHEPSHED,
LEICESTERSHIRE LE12 9NH, ENGLAND
TELEPHONE: 509 600220 FAX: 509 508795

BUYING & SELLING

AUSTRALIA — JAPAN

	B1 GALLERY	B2 RETAIL	B3 WHOLESALE	B4 MAILORDER	B5 FINE ART	B6 CUSTOM	B7 MARKETING CONSULTANT	B8 POINT-OF-PURCHASE DISPLAYS	B9 CUSTOM
AUSTRALIA									
er Light Expressions Pty. LTD	✓	✓	✓		✓	✓	✓	✓	✓
art Pty. Ltd.	✓	✓	✓						
w Dimension Holographics		✓	✓		✓	✓		✓	
BELGIUM									
logram Europe sprl.									
ger Graphic Technology		✓					✓		
CANADA									
nge Research Holographics	✓		✓		✓				
lo Laser Tech	✓								
lomagic Inc.									
OU Inc.	✓	✓	✓	✓	✓	✓	✓	✓	✓
erference Hologram Gallery	✓	✓	✓	✓	✓	✓		✓	✓
ser Holographics, Inc.		✓			✓				
ral University (Ave.des Laurentides)					✓				
ht Construction, Inc.					✓		✓	✓	✓
nna J. Roa									
lone Holographie Corp.	✓	✓	✓		✓				
DENMARK									
w Dimensions Exhibition	✓								
FRANCE									
see De L'Holographie	✓	✓							
see Francais De l'Holographie	✓	✓	✓						
FINLAND									
olografia Galleria	✓	✓	✓						
ISRAEL									
ird Dimension Ltd.	✓	✓	✓						
ITALY									
olographic Service							✓		
lotron SRL						✓	✓		
tamorfosi Olografia Italia			✓						
JAPAN									
lo ARP	✓				✓				
miko Shiozaki					✓				

BUYING & SELLING

MEXICO — UNITED KINGDOM	B1 GALLERY	B2 RETAIL	B3 WHOLESALE	B4 MAILORDER	B5 FINE ART	B6 CUSTOM	B7 MARKETING CONSULTANT	B8 POINT-OF-PURCHASE DISPLAYS	B9 CUSTOM
MEXICO									
...ologramas De Mexico		✓	✓			✓			
NETHERLANDS									
...utch Holographic Laboratory		✓	✓						
...ue Sera Sera			✓						
SAUDI ARABIA									
...onders of Holography Gallery	✓								
SPAIN									
...olos S.A.							✓		
...olotek							✓		
SWEDEN									
...olo Media AB/Hologram Museum	✓	✓	✓	✓	✓	✓	✓	✓	✓
...pectrogon AB							✓		
...vede Holoprint			✓				✓		
SWITZERLAND									
...olodesign Studies							✓		
...olos Art Galerie	✓	✓	✓						
...aserart Ford AG		✓	✓						
...allerie Fur Hologrammine	✓								
UNITED KINGDOM									
...H. Prismatic, Inc.	✓	✓	✓	✓	✓		✓		
...mazing World of Holograms			✓	✓				✓	✓
...pplied Holographics			✓						
...argaret Benyon					✓				
...ff Blyth									
...ambridge Consultants Ltd.									
...ar East Holographics Ltd.			✓						
...ologram One		✓							
...ologram Roadshow						✓	✓	✓	
...olotec PLC						✓	✓		
...olovision						✓	✓		
...ndrew Laczynski							✓		
...aser Light Image		✓	✓	✓	✓				
...ght Impressions Europe, Ltd	✓	✓	✓		✓				✓

BUYING & SELLING

	B1	B2	B3	B4	B5	B6	B7	B8	B9
	GALLERY	RETAIL	WHOLESALE	MAILORDER	FINE ART	CUSTOM	MARKETING CONSULTANT	POINT-OF-PURCHASE DISPLAYS	CUSTOM
ITED KINGDOM (continued) —									
nday Spatial Imaging	✓	✓	✓	✓	✓				
Clear Imports Ltd.									
id Pizzanelli							✓		
nium Technology Ltd.	✓	✓	✓						
3 Holograms Ltd.		✓	✓✓						
ITED STATES OF AMERICA									
2000	✓	✓	✓✓	✓			✓✓	✓	✓
otech Inc.									
. Prismatic Inc.	✓	✓	✓	✓					
s Lightworks			✓	✓	✓✓				✓✓
it Studio		✓	✓	✓		✓			✓
other Dimension									
lied Holographics, Corp.			✓		✓				
gliography	✓		✓						
s, Barefoot and Associates			✓✓✓✓						
lie Berkhout							✓		
l Gallery					✓		✓		
us Brill	✓✓	✓✓	✓✓		✓✓	✓✓	✓✓	✓✓	✓✓
din-Silver Holography (Boston)									
din-Silver Holog. (Brookline)					✓	✓	✓	✓	✓
ssey Image Center									
rry Optical Company	✓								
cago Museum of Holography									
urn Corporation	✓				✓		✓		
Company		✓	✓				✓✓		
tman Kodak Company									
nund Scientific Co.									
sive Image	✓								
sive Image									
er Scientific									
hur David Fornari	✓	✓	✓	✓	✓✓	✓	✓✓✓	✓✓✓	✓✓✓
ographic Studios									
M	✓	✓	✓	✓	✓	✓			
laxis Corporation									

BUYING & SELLING

S.A. (continued)	B1 GALLERY	B2 RETAIL	B3 WHOLESALE	B4 MAILORDER	B5 FINE ART	B6 CUSTOM	B7 MARKETING CONSULTANT	B8 POINT-OF-PURCHASE DISPLAYS	B9 CUSTOM
Hologram World	✓	✓	✓	✓	✓				
Holographic Concepts							✓	✓	
Holographic Design, Inc.			✓				✓	✓	
Holographic Design Systems, In	✓	✓	✓	✓			✓	✓	✓
Holographic Visions	✓						✓		
Holography News							✓		
Holography Workshops	✓	✓	✓	✓	✓	✓		✓	✓
Holos Gallery	✓	✓	✓	✓		✓		✓	
Holo/Source Corporation			✓				✓	✓	
Holo-Spectra		✓	✓	✓	✓		✓	✓	✓
Holosystems, Inc.									
Imaging & Design			✓						
Integraf							✓		
Interaction in Laser Light							✓		
James River Products							✓		
Just Books and Holographics							✓		
Jon Kaufman Photography					✓		✓		
L.A.S.E.R. Co.		✓		✓	✓	✓	✓		✓
Laser Dreams	✓	✓	✓	✓	✓				✓
Laser Focus	✓	✓					✓		
Laser Light Designs	✓	✓	✓	✓		✓		✓	
Les Heures Ltd.		✓	✓	✓	✓	✓	✓		✓
Let There Be Neon			✓				✓		
Light Harmonics Inc.									
Light Impressions Inc.							✓	✓	
Light Wave	✓			✓	✓	✓	✓	✓	✓
McCain Marketing									
Media Interface, Ltd							✓		
Mind's Eye: Holog. Consultants					✓		✓		
William Molteni	✓			✓			✓		
Newport Research Corporation							✓		
New York Holographic Lab.			✓				✓		
...rtson Inc.									
connaissance									

BUYING & SELLING

...S.A. (continued) — WEST GERMANY

	B1 GALLERY	B2 RETAIL	B3 WHOLESALE	B4 MAILORDER	B5 FINE ART	B6 CUSTOM	B7 MARKETING CONSULTANT	B8 POINT-OF-PURCHASE DISPLAYS	B9 CUSTOM
...hool of Holography/Chicago	✓	✓	✓	✓	✓				
...bert Sherwood Holog. Design		✓	✓			✓	✓	✓	✓
...ace Age Designs							✓	✓	
...nchronicity Holograms									
...ird Dimension Arts, LTD		✓	✓	✓					
...ree-D Light Gallery	✓	✓		✓	✓				
...avefront Reconstruction Compa							✓		
...ave Mechanics							✓		
...illamette Science & Technology									
...onderlight Gallery	✓	✓	✓	✓	✓				
WEST GERMANY									
...S Holographie-Galeria GmbH	✓	✓							
...OL 3 Galerie fur Holographie	✓	✓							
...OLO GmbH	✓	✓	✓	✓	✓		✓		✓
...olographie Labor	✓	✓			✓				
...olographie Galerie		✓							
...useum fur Holographie	✓	✓							
LATE ENTRY: USA									
...olography Institute	✓	✓	✓	✓	✓		✓	✓	✓
...K Gold Purchasers Inc.							✓		

BUYING & SELLING

B1=Gallery	**B4**=Mailorder	**B7**=Marketing Consultant
B2=Retail	**B5**=Fine Art Holograms	**B8**=Point of Purchase Displays
B3=Fine Art Ltd Editions	**B6**=Custom Packaging	**B9**=Custom Lighting

AUSTRALIA

Laser Light Expressions Pty. Ltd
G 12 Denawen Ave. Castle Grove
P.O. Box 23 Chatswood 2057
N.S.W., Australia
Telephone: (61)(612) 4072896
ContactJ.Tobin, Rosie
FAX: (61)(612)4177247
Categories: Al A2 A3 A4 A5 A6 A7 A8 A9 AI0 B2
B3 B5 B6 B7 B8 B9 C4

Lazart Pty. Ltd.
22 Erina Valley Road
Erina
New South Wales
2250, Australia
Telephone: (61)(43) 676245
Contact: Roslyn Wilson
Categories: A7 A8 Bl B2 B3

New Dimension Holographies
65-72 Pier One
Hickson Road
Sydney, New South Wales 2000
Australia
Telephone: (61) (02) 276063
Contact: Tony Butteriss
Categories: Bl B2 B3 B5 B6 B8 C7

BELGIUM

Hologram Europe sprl.
Avenue Voltaire 137
1030 Brussels, Belgium
Telephone: (32) (2) 242 7284
Contact J. B. Boulton
Categories: Al B2

Jaeger Graphic Technology
20 Avenue des Desirs
B-1140 Brussels, Belgium
Telephone: (32) (2) 7359551
Contact In. Jaeger
Telex: 23027 JAEGER B
FAX: (32) (2) 7331035
Categories: A7

CANADA

Fringe Research Holographics
1179A King Street, West
SuiteOOS
Toronto, Ontario
Canada M6K 3C5
Telephone: (416) 535 2323
Contact: Michael Sowdon, Dir.
Categories: A2 A3 A7 A8 A11 B1 B5 C4 C8 C10

Holo Laser Tech
7 Fraser Avenue #16
Toronto, Ontario
Canada, M6K IV?
Telephone: (1) (416) 638-7991
Contact Glen Stradz
Categories: A1 B3

Holomagie Inc.
91717thAvenueSW
Calgary, Alberta
Canada TIT OA4
Telephone: (1) (403) 229-0069
Contact Linda Parkes
Categories: Bl B2B5

IBOVInc.
CP214
Cap-de-Ia-Madeleine
Quebec, Canada G8T 7W2
Telephone: (1) (819) 295 5~~
Categories: A1 A2 A3 A4 A7 A8 A9 B2 B3 B4 B5
B6 B8 B9

Interference Hologram Gallery
1179A King Street West
Toronto, Ontario
Canada M6K 3C5
Telephone: (1) (416) 535 2323
Categories: B1 B2 B5

A.H. PRISMATIC

presents the

HOLOGRAM CENTRE

A Captivating In-Store Display for All Our Merchandise

The Best Products . . . The Best Designs . . . The Best Packaging . . . The Best Prices

A.H. PRISMATIC

Enquiries in the U.S.A. to:
A. H. PRISMATIC INC.
285 West Broadway
New York, New York 10013. U.S.A.
Tel: (212) 219 0440
Telex: 6973539 PRISM NY

All enquiries (excluding U.S.A.) to:
A. H. PRISMATIC LTD.
New England House, New England Street
Brighton, Sussex BN1 4GH. England
Tel: (0273) 686966
Telex: 877668 PRISM G

Pictured above: Floor Standing Unit displaying Film Holograms, Laser Jewellery, Laser Discs, Laser Spex, Hologram Jigsaws, Stickers and Boxes.

BUYING & SELLING

B1=Gallery   **B4**=Mailorder   **B7**=Marketing Consultant
B2=Retail   **B5**=Fine Art Holograms   **B8**=Point of Purchase Displays
B3=Fine Art Ltd Editions   **B6**=Custom Packaging   **B9**=Custom Lighting

Laser Holographies, Inc.
1179 King St. West
Unit III
Toronto, Ontario
Canada M6K 3C5
Telephone: (1) (416) 531 4656
Contact: Charles Demicher
Telex. 7601314 (UI)
FAX: (1) (416) 530 1594
Categories: A1 A6 Bl B3 B4 B5 B6 B7 B8 B9

Laval University
1145 Ave. des Laurentides
Quebec City, Quebec
Canada GIS 3C2
Telephone: (1) (418) 6872985
Contact. Marie Cossette
Categories: A1 A2 A3 B5 C3 C4 C5 C12

Light Construction, Inc.
2154 Dundas Street West
Suite #503
Toronto, Ontario
Canada M6R IX3
Telephone: (1) (416) 5334692
FAX: (1)(416) 5330572
Categories: Al A2 A 7 A8 A10 B5 C12

WannaJ.Roa
2450 Lancaster Road Unit 36
Ottawa, Ontario
Canada, KlB 5N3
Telephone: (1) (613) 521 2557
Categories: B7

BUYING & SELLING

B1=Gallery	**B4**=Mailorder	**B7**=Marketing Consultant
B2=Retail	**B5**=Fine Art Holograms	**B8**=Point of Purchase Displays
B3=Fine Art Ltd Editions	**B6**=Custom Packaging	**B9**=Custom Lighting

Trilone Holographie Corp.
4200 Blvd. St Laurent
Montreal, Quebec
H2W 2R2 Canada
Telephone: (1) (514) 8456992
Contact: Gerard Allon
FAX: (1)(514) 8498706
Categories: Al A2 A3 A4 A7 A8 A9 AI0 All B3 B5
B8 B9

DENMARK

New Dimensions Exhibition
The H.C. Andersen Castle
H.C. Andersens Blvd. 22
1553 Copenhagen V
Denmark
Telephone: (45)(1) 131 713
Categories: Bl B2

FRANCE

Musee De L'Holographie
Forum des Halles 15 A 21
Grand Balcon - 75001 Paris
France
Telephone: (33) (14) 296 9683
Categories: B1 B2 C1

Musee Francais De l'Holographie
4 Rue Beaubourg
75004 Paris
France
Telephone: (33) (14) 2771512
Categories: Bl B2

BUYING & SELLING

B1=Gallery
B2=Retail
B3=Fine Art Ltd Editions

B4=Mailorder
B5=Fine Art Holograms
B6=Custom Packaging

B7=Marketing Consultant
B8=Point of Purchase Displays
B9=Custom Lighting

FINLAND

Holografia Galleria
Jaakonkatu 3 (2nd floor)
00100 Helsinki
Finland
Telephone: (358) (0) 6941 909
Categories: B1 B2

ISRAEL

Third Dimension Ltd.
Dizengof Centre
Tel-Aviv
Israel
Telephone: (972) (3) 287264
Categories: Bl B2B3

ITALY

Holographic Service
10 Via Civerchio
20159 Milan
Italy
Telephone: (39) (2) 688 7067
Categories: B3 B6 B7

Holotron SRL
Via T olstoi 46
20146 Milan
Italy
Telephone: (39) (2) 479 697
Categories: B 7

Metamorfosi Olografia Italia
Via Lecco 6
20124 Milan
Italy
Telephone: (39) (2) 27993
Categories: B3

JAPAN

HoloARP
Kamakura Inc.
7-10-8 Ginza Chuo-Ku
Tokyo
Japan
Telephone: (81) (03) 5748307
Contact Yumiko Shiozaki
FAX: (81) (03) 5748377
Categories: A1 A2 A3 A4 A6 B1 B3 B5 C3 C6 C9

Yumiko Shiozaki
7-5-18-Ryoke Urawa
Saitama
Japan
Telephone: (81) (0488) 31 7723
Categories: A2 B5 C3 C5 C7 C8 C9 CI0

MEXICO

Hologramas De Mexico
PINO 343, Local 3
COL. ST A. MA IA RIBERA
06400 Mexico, D.F.
Mexico
Telephone: (52)(905) 5479046
Contact: Dan Iieberman
FAX: (52) (905) 5474084
Categories: A1 A2 A4 A5 A6 A8 A9 A12 B2 B3 B6
C10

NETHERLANDS

Dutch Holographic Laboratory
Kanaaldyk Noord 61
Eindhoven 5642 JA
The Netherlands
Telephone: (31)(40) 817250
Contact: Walter Spierings, Dir.
FAX: (31) (40) 814865
Categories: A1 A3 A4 AS A6 A7 A8 A9 A10 B3 D1

BUYING & SELLING

B1=Gallery	**B4**=Mailorder	**B7**=Marketing Consultant
B2=Retail	**B5**=Fine Art Holograms	**B8**=Point of Purchase Displays
B3=Fine Art Ltd Editions	**B6**=Custom Packaging	**B9**=Custom Lighting

Que Sera Sera
P.O. Box 29
9700 AA Groningen
The Netherlands
Telephone: (31) (050) 140417
Contact: H.T. Vogd
FAX: (31)(050) 144142
Categories: A12 B3 C4 D3 D4

SAUDI ARABIA

Wonders of Holography Gallery
P.O. Box 1244
Jeddah 21431
Saudi Arabia
Telephone: (966) (2) 688 5208
Categories: B 1 B2

SPAIN

HoiosSA
Calle Gobernador Viejo 30 Bajo
Valencia 46003
Spain
Telephone: (34) (96) 3333013
Categories: B 7

Holotek
Granda 47 Siero
Asturias
Spain
Telephone: (34) (985) 793526
Contact: Julio Ruiz Garcia
Categories: Al B 7

SWEDEN

Holo Media AB/Hologram Museum
Box 45012
Drottninggatan 100
10430 Stockholm, Sweden
Telephone: (46) (08) 105465
Contact. Mona Forsberg
FAX: (46)(08) 107638
Categories: Al A6 A11 A12 B1 B2 B3 B4 B5 B6 B7 B8 B9
C1 C5 C10

Spectrogon AB
Box 2076
S-18302 Taby
Sweden
Telephone: (46) (8) 7580980
Categories: B7

Swede Holoprint
Duvhoksgatan 6 A
Malmo 21460
Sweden
Telephone: (46)(40) 89821
Contact: Bjorn Wahlberg
Categories: B7

SWITZERLAND

Holodesign Studies
Rebenstrasse 20
CH4125 Riehen
Switzerland
Telephone: (41) (61) 672342
Categories: B 7

Holos Art Galerie
4 Place Grenus
1201 Geneva
Switzerland
Telephone: (41) (22) 325 191
Categories: Bl B2

Laserart Ford AG
Rebenstrasse 20
CH 4125 Riehen/Basel
Switzerland
Telephone: (41) (61) 672 343
Categories: Al A8 B2 B3

Gallerie Fur Hologrammine
H. Gutekunst
Via Principale 30
CH - 7649 Castesegua
Switzerland
Telephone: (41)(082) 41718
Categories: B1

BUYING & SELLING

B1=Gallery	**B4**=Mailorder	**B7**=Marketing Consultant
B2=Retail	**B5**=Fine Art Holograms	**B8**=Point of Purchase Displays
B3=Fine Art Ltd Editions	**B6**=Custom Packaging	**B9**=Custom Lighting

UNITED KINGDOM

A.H. Prismatic, Inc.
New England House
New England Street, Brighton
BN1 4GH England
United Kingdom
Telephone: (44) (273) 686966
Contact. Simon Lee
FAX: (44) (273) 676692
Categories: A8 B3

Amazing World of Holograms
The Gallery, Corrigans Arcade
Foreshore Road, South Bay
Scarborough, North Yorkshire
England
Telephone: (44) (0723) 354090
Contact: Carl Racey
Categories: B1 B2 B3 B4 B5 B8 B9

Applied Holographics
Braxted Park, Witham
Essex, CM8 3XB
England
United Kingdom
Telephone: (44) (621) 893030
Categories: A7A8B3

Margaret Benyon
Holography Studio
40 Springdale Avenue
Broadstone, Dorset
BH18 9EU, England
Telephone: (0202) 698067
Contact: Margaret Benyon
Categories: A2 A3 A7 A8 B5 C4

Jeff Blyth
7 Bath Street
Brighton, East Sussex
BN1 3TB, England
United Kingdom
Telephone: (44) (273) 202069
Categories: B7

Cambridge Consultants Ltd.
Science Park
Milton Road, Cambridge
CB4 4DW, England
United Kingdom
Telephone: (44) (223) 358855
Categories: B7

Far East Holographics Ltd.
220 Spring Bank
Hull, North Humberside
England HU3 1LU
United Kingdom
Categories: B3

Hologram One
39 Pyrcoft Road
Chertsey, Surrey
England KT16 9HT
United Kingdom
Telephone: (44) (9328) 64899
Categories: A1A8B3

Hologram Roadshow
Longleatr House
Warminister, Wiltshire
12 Queen Square
Bath, Avon BA1 1 WU England
Telephone: (44) (225) 339333
Categories: B2

HoiotecPLC
7 Cameron Road
Seven Kings, Ilford
Essex, 1 G3 BLG
England, United Kingdom
Telephone: (44) (1) 5978004
Categories: Al B6 B7 B8

BUYING & SELLING

B1=Gallery	**B4**=Mailorder	**B7**=Marketing Consultant
B2=Retail	**B5**=Fine Art Holograms	**B8**=Point of Purchase Displays
B3=Fine Art Ltd Editions	**B6**=Custom Packaging	**B9**=Custom Lighting

Holo VIsion
43 Pall Mall
London SWIY 55G
England
United Kingdom
Telephone: (44)(1) 8395622
Categories: B 7

Andrew Laczynski
22 Woodstock Ave.
London W139UG
England
United Kingdom
Telephone: (44)(1) 8405101
Categories: Al B6

Laser Light Image
101 Spring Bank
Hull, HU3 lBH
England
United Kingdom
Telephone: (44) (0482) 26744
Contact: Carl Racey
FAX: (44) (0482) 492286
Categories: Bl B2 B3 B4 B5 B9

Light Impressions Europe, Ltd
Attn: Kenneth Harris
12 Mole Business Park, Off Sta
Leatherhead, Surrey KT22 7AQ
England, United Kingdom
Telephone: (44) (372) 386 677
Categories: Al A6 B3 B5

Munday Spatial Imaging
39 Pyrcroft Road
Chertsey, Surrey
KTl6 9HT, England, U.K.
Telephone: (44) (0932) 564 899
Contact: Rob Munday
Categories: Al A2 A3 A4 A7 A8 All Bl B2 B3 B4
B5

New Clear Imports Ltd.
27 Burrard Street
St Helier
Jersey, Channel Islands
England
Telephone: (44)(534) 0630614
Categories: B3

David Pizzanelli
4 Maculay Road
London, SW40QX
England
United Kingdom
Telephone: (44)(1) 6271140
Categories: Al A2 A3 A4 A6 A7 A8 A9 B7 C9

Premium Technology Ltd.
Holomart
9 Brunswick Centre
London WCIN lAF
England
Telephone: (44) (1) 8330998
Categories: Bl B2 B3

See-3 Holograms Ltd.
13 Bovingdon Road
London SW62AP
England
United Kingdom
Telephone: (44) (1) 736 0076
Categories: B3

UNITED STATES OF AMERICA

AD 2000
946 STATE STREET
NEW HAVEN, CT
06511
USA
Telephone: (1) (800) 334 4633
Contact: Jeffrey Levine
Categories: Al A4 A6 A7 A8 A9 A12 B7 B8 B9

BUYING & SELLING

B1=Gallery   **B4**=Mailorder   **B7**=Marketing Consultant
B2=Retail   **B5**=Fine Art Holograms   **B8**=Point of Purchase Displays
B3=Fine Art Ltd Editions   **B6**=Custom Packaging   **B9**=Custom Lighting

Aerotech Inc.
Electro Optical Division
101 Zeta Drive
Pittsburgh, PA
15238
Telephone: (1) (412) 9637470
Categories: B7

A.H. Prismatic Inc.
285 West Broadway
New York, NY
10013
Telephone: (1) (212)2190440
Contact: Andrew Meehan
FAX: (1)(212) 2190443
Categories: A8 Bl B3 B4

Aites Lightworks
2148 North 86th Street
Seattle, WA
98103
Telephone: (1) (206) 5265752
Contact: Edward Aites
Categories: A2 A3 A 7 A8 B2 B5

Anait Studio
1685 Fernald Point Lane
Santa Barbara, Calif.
93108
Telephone: (1) (805) 9695666
Contact Anait
Categories: A2 A3 A11 B5 C4 C12

Another Dimension
948 State Street
New Haven, cr
06511
USA
Telephone: (1) (203) 6246405
Telephone: (1)(800) 3344633
Contact. Peter Scheir
FAX: (1)(203) 6241780
Categories: BI B2 B3 B4 B6 B9

Applied Holographics, Corp.
1530 Progress Road
P.O. Box 8300
Fort Wayne, IN
46898
Telephone: (1)(219) 4846081
Contact Richard W. Gipp
FAX: (1)(219) 4848611
Categories: Al A4 A6 A 7 A8 All B8 B9

Artigliography
7130 Mohawk West Drive
Indianapolis, IN
46236
USA
Telephone: (1)(317) 8230069
Contact. Kerry S. Brown
Categories: A2 A 7 A8 B3 B5 C3 C6 C8 Cll

Bliss, Barefoot and Associates
500 Fifth Avenue
New York, NY
10110
Telephone: (1) (212) 8401661
Contact: Paul D. Barefoot
FAX: (1) (212) 8401663
Categories: AI B7

Rudie Berkhout
223 West 21st Street, Apt. B
New York, NY
10011
Telephone: (1) (212) 2557500
Categories: A2 A3 B5

Brill Gallery
77 Main Street
Cold Spring, NY
10516
Telephone: (1) (914) 265 2326
Categories: B1

BUYING & SELLING

B1=Gallery	**B4**=Mailorder	**B7**=Marketing Consultant
B2=Retail	**B5**=Fine Art Holograms	**B8**=Point of Purchase Displays
B3=Fine Art Ltd Editions	**B6**=Custom Packaging	**B9**=Custom Lighting

Louis Brill
2428 Judah St
San Francisco, CA
94122
USA
Telephone: (1) (415) 664 0694
Categories: B 7

Casdin-Silver Holography
51 Melcher Street
Studio 501
Boston, MA 02210
Telephone: (1)(617)4234717
Contact: Harriet Casdin-Silver
Categories: Al A2 A3 A4 AS A7 A8 A9 AI0 All B2
B3 B5 B6 B8 B9 C2 C5 C6 C7 C9 ClO Cll

Casdin-Silver Holography
99 Pond Avenue
SuiteD403
Brookline,MA
02146
Telephone: (1) (617) 7396869
Contact: Harriet Casdin-Silver
Categories: Al A2 A3 A4 A5 A7 A8 A9 AI0 All B2
B3 B5 B6 B8 B9 C2 C5 C6 c7 C8 CI0 Cll C12

Odyssey Image Center
8853 Sunset Boulevard
Hollywood, CA
90096
Telephone:(1) (213) 6520983
Categories: B1

Cherry Optical Company
2047 Blucher Valley Road
Sebastopol, CA
95472
Telephone: (1)(707) 8237171
Contact: G.Cherry IN.Gorglione
Categories: Al A2 A3 A4 A7 A8 A9 A10 B3 B4 B7
C2 C3 C6 C10

Chicago Museum of Holography
1134 West Washington Boulevard
Chicago,IL
00007
USA
Telephone: (1) (312) 2261007
Contact: L. Billings
Categories: Bl B2 B3 B5 Cl C2 C7 CI0 Cll C12

Coburn Corporation
1650 Corporate Road West
Lakewood, NJ
08701
Telephone: (1) (201)367 5511
Contact: Joseph Coburn III
Telex. 132438
FAX: (1) (201) 3672908
Categories: A6 B3

DZCompany
181 Mayhew Way Suite E
P.O. Box 5047-T
Walnut Creek, CA 94596
Telephone: (1) (415) 9354656
Contact: Dan Cifelli
FAX: (1)(415) 9354660
Categories: Al A4 A6 B3 B6 B7 B8

Eastman Kodak Co
Scientific & Technical Photography
343 State Street
Rochester, NY 14650
Telephone: (1) (716) 7244000
Contact: RD. Anwyl
Categories: B7

Edmund Scientific Co.
101 East Gloucester Pike
Barrington, NJ 08007
Telephone: (1)(009) 5478900
Categories: B2 C8

BUYING & SELLING

B1=Gallery	**B4**=Mailorder	**B7**=Marketing Consultant
B2=Retail	**B5**=Fine Art Holograms	**B8**=Point of Purchase Displays
B3=Fine Art Ltd Editions	**B6**=Custom Packaging	**B9**=Custom Lighting

Elusive Image
603 Munger Street # 316
Dallas, TX
75202
Telephone: (1)(214) 521 9309
Contact. Fred Wtlbur
Categories: B1

Elusive Image
4514 Travis Street#1l4
Dallas, TX
75205
Telephone: (1)(214) 5219309
Contact. Fred Wtlbur
Categories: B1

Fisher Scientific
E.In. Division
4901 West Lemoyne Avneue
Chicago,IL
USA
Telephone: (1) (312) 3782208
Categories: B2

Arthur David Fornari
813 Eighth Avenue
Brooklyn, NY
11215
Telephone: (1) (71) 965 3956
Categories: A2 A 7 A8 B5

GPM
4165 Apalogen Road
Philadelphia, PA
19144
USA
Telephone: (1) (215) 8494049
Categories: Al A4 A6 A7 A8 A9 AIO All AI2 B7 B8
B9

Holaxis Corporation
968 Farmington Avenue
Hartford, CT
06107
USA
Telephone: (1)(203)2322030
Contact Martin Berson
Categories: Al A4 A5 A6 A7 A8 A9 AIO All AI2 BI
B3 B4 B5 B6 B7 B8 B9

Hologram World
12121/2 Dixon Boulevard
Cocoa,FL
32922
USA
Telephone: (1) (407) 631 3615
Contact Susan K Harrison
Categories: BI B2 B3 B4 B5

Holographic Concepts
711 East 13th Street
Houston, TX
77008
USA
Telephone: (1) (713) 861 2865
Categories: B7

Holographic Design, Inc.
1084 North Delaware Avenue
Philadelphia, PA
19125
USA
Telephone: (1) (215) 4259220
Contact D .Miller
Categories: Al A4 A6 AI2 B3 B6 B7 B8 C7

Holographic Design Systems, Inc
1134 West Washington Blvd.
Chicago,IL
60607
USA
Telephone: (1)(312) 829 2292
Contact Robert Billings
Categories: Al A2 A3 A4 A5 A6 A7 A8 A9 AIO B7
B8

BUYING & SELLING

B1=Gallery   **B4**=Mailorder   **B7**=Marketing Consultant
B2=Retail   **B5**=Fine Art Holograms   **B8**=Point of Purchase Displays
B3=Fine Art Ltd Editions   **B6**=Custom Packaging   **B9**=Custom Lighting

Holographic Studios
240 East 26th Street
New York. NY 1 0010
Contact. Jason Sapan
Telex. (1)(212) 6869397
FAX: (1)(212) 481 S645
Categories: A1 A2 A3 A4 A5 A6 A7 AS A9 A12 B1
B2 B3 B4 B5 B6 B7 BS B9 C4 C12

Holographic Visions
300 South Grand Avenue
Los Angeles, CA
90071
Telephone: (1)(213) 6877171
Contact. Mark Merrill
Categories: B1 B2 B3 B4 B5 B6 B7 BS B9 C1 C2 CS
C11

Holography News
3932 Mckinley Street, N.W.
Washington, D.C.
20015
Contact. L Kontnick
Categories: Al B7 C12

Holography Workshops
Lake Forest College
Sheridan and Collge Road
Lake Forest, IL
60045
Telephone: (1) (312) 2343100
Contact. Tung H,Jeong
Categories: Al A2 A4 A7 AS A9 Ala All A12 Bl Cl
C2 C3 C4 C5 C6 C7 C8 C9 ClO C11 C12

Holos Gallery
1792 Haight Street
San Francisco, CA
94117
USA
Telephone: (1) (415) 861 0234
Contact Gary Zellerbach, A.Rhody, Sales Mgr.
FAX. (1)(415) S611563
Categories: Bl B2 B3 B4 B5 B6 B7 B8

Holo/Source Corporation
21S00 Melrose Avenue
Suite 7
Southfield, MI 48075
Telephone: (1) (313) 3550412
Cantact: L. Lacey, B. Seidel
FAX: (1) (313) 3550437
Categories: B3 B4 B6 B7 B8

HoloSpectra
7742-B Gloria Avenue
VanNuys,CA
91406
Telephone: (1) (SIS) 9942577
Contact. W Arkin
FAX: (1) (SIS) 994 4709
Categories: A1 A4 A5 A6 A 7 AS A9 A12 B3 BS D2
D3 D4 D5 D6 D7 DS D9 D10 D11 D12

Holosystems, Inc.
P.O. Box 6810
Ithaca,NY
14851-6810
USA
Telephone: (1) (607) 273 11S7
Contact. Jonathan Back
Categories: B3 B4 B7

Imaging & Design
11 01 Ransom Road
Grand Island, NY
14072-1459
Telephone: (1) (716) 7737272
Contact. Keith Allen
Categories: Al A4 A6 A7 AS A9 Ala All B7 BS B9

Integra£'
745 North Waukegan Road
Lake Forest, IL
60045
Telephone: (1) (312) 2343756
Contact. Anna Wong
Categories: B2 B3 B4 B7 C3 C6 C7 Dl D3 D6

BUYING & SELLING

B1=Gallery   **B4**=Mailorder   **B7**=Marketing Consultant
B2=Retail   **B5**=Fine Art Holograms   **B8**=Point of Purchase Displays
B3=Fine Art Ltd Editions   **B6**=Custom Packaging   **B9**=Custom Lighting

Interaction in Laser Light
5326 Sunset Blvd.
Los Angeles, CA
90027
USA
Telephone: (1)213) 466 5767
Categories: B 7

James River Products
5420 Distributor Dr.
Richmond, VA
23225
USA
Telephone: (1)(804) 2339145
Categories: B7

Just Books and Holographies
Grogan's Mill Village Center
2250 Buckthome
The Woodlands, TX
77380
Telephone: (1) (713) 3676258
Categories: B2

John Kaufman Photography
P.O. Box 477
Point Reyes Station, CA 94956
Telephone: (1) (415) 6631216
Cantact John Kaufman
Categories: A2 A3 A8 A10 B5 C4

L.AS.E.R. Co.
1900 Gore Drive
Haymarket, VA
22000
Telephone: (1) (703) 7542526
Contact. Jim Bowman
Categories: A2 A3 A7 A8 A9 B2 B5 B9 C4 C12

Laser Dreams
Attn: Nancy J. Gorglione
P.O. Box 326
Forestville, CA
95436
Telephone: (1)(707) 8236104
Categories: A2 A3 A4 A7 A8 A9 A10 B2 B3 B6 B9 C1

Laser Focus
1 Technology Park Drive
Westford, MA
01886
Telephone: (1) (508) 6920700
Categories: B4

Laser Light Designs
2412 Kennedy Way
Antioch, CA
94509
Telephone: (1) (415) 7543144
Contact: Michael Malott
Categories: Bl B2 B3 B4 B7 B8 CI0

Les Heures Ltd.
P.O. Box 1437
Canal Street Station
New York, NY 10013
Telephone: (1) (212) 869 3050
Categories: B2

Let There Be Neon
P.O. Box 337
Canal Street Station
New York, NY 10013
Telephone: (1)(212) 2264883
Categories: Bl

Light Harmonics Inc.
93 Lake Shore Drive
Oakland, r.rr
07436
Telephone: (1) (201) 3378868
Contact Jonathan Klempner
FAX: (1)(201 4329542
Categories: Al A2 A3 A5 A6 A7 A8 A9 AI0 All B2
B3 B5 B7 B9 C3 C4 C6 c7

Light Impressions Inc.
149 B Josephine Street
Santa Cruz, CA
95000
Telephone: (1) (408) 458 1991
Contact: Kathryn S. Long, Mgr.
FAX: (1)(408) 4583338
Categories: Al AS A6 B2 B3 B4 B6 B7 C8

BUYING & SELLING

B1=Gallery
B2=Retail
B3=Fine Art Ltd Editions

B4=Mailorder
B5=Fine Art Holograms
B6=Custom Packaging

B7=Marketing Consultant
B8=Point of Purchase Displays
B9=Custom Lighting

lightWave
D-238
Woodfield Mall
&haumburg,IL
60173
Telephone: (1) (312) 204 5344
Cantact: Gordon Savage
Categories: BI B2 B5 B8 B9

McCain Marketing
10962 North Wauwatosa Road
76W
Mequon, WI
53092
Telephone: (1)(414) 2424023
Contact: Richard McCain
Categories: B4 B7 C6 C8 CI2

Media Interface, Ltd.
167 Garfield Place
Brooklyn, NY
11215
Telephone: (1) (718) 7884012
Contact Ronald R. Erickson
Categories: A1 A4 A9 B9 C3 C6

Mind's Eye: Holographic Consultants
17329 Zola Street
Granada Hills, CA
91344
USA
Telephone: (1) (818) 360 6023
Contact: Stephen Roth
Categories: B 7

William Molteni
265 Elm Street
Somerville, MA
02144
Telephone: (1)(617) 6668144
Categories: B 7

Newport Research Corporation
18235 Mount Baldy Circle
Fountain Valley, CA
92708
USA
Telephone: (1)(714) 9639811
Contact Technical Suppt
Categories: B4 Dl D4 D5 D9 D 12

New York Holographic Laboratory
P.O.Box 20391
Tomkins Square Station
New York, NY
10000
Telephone: (1)(212) 2549774
Contact: Dan &hweitzer
FAX (1) (212) 6741007
Categories: Al A2 A3 A6 A7 A8 BI B5 C3 C4

Portson Inc.
9201 Quivira
Overland Park, KS
66215-3905
Telephone: (1) (913) 4927010
Contact: Jill Jarvis
FAX (1) (913) 4927099
Categories: Al A4 A 7 A8 A10 A12 B3

Reconnaissance
3932 McKinley Street, N.W.
Washington, D.C.
20015
USA
Cantact: L. Kontnick
FAX: (1)(703) 764-6398
Categories: Al B7 CI2

School of Holography/Chicago
1134 West Washington Boulevard
Chicago,IL
60607
Telephone: (1) (312) 226 1007
Contact: Loren Billings, Dir.
Categories: B1 B2 B3 B4 B5 B7 C1 C2 C3 C4 C5 C6
C7 C8 C9 C10 C11 C12 D1 D6

BUYING & SELLING

B1=Gallery   **B4**=Mailorder   **B7**=Marketing Consultant
B2=Retail   **B5**=Fine Art Holograms   **B8**=Point of Purchase Displays
B3=Fine Art Ltd Editions **B6**=Custom Packaging   **B9**=Custom Lighting

Robert Sherwood Holog. Design
400 West Erie Street
Chicago,IL
60610
USA
Teleplwne: (1) (312) 944 0784
Contact: K.Kellison,R.Sherwood,
FAX: (1)(215) 4259221
Categories: Al A4 A6 AI2 B6 B7 B8 B9 C7 CII

Space Age Designs
P.O. Box 72
Carversville, PA
18913
Teleplwne: (1) (215) 2978490
Contact: Valli Rothaus
Categories: A2 A3 AI2 B3 B7

Synchronicity Holograms
Box 4235
Lincolnville, ME
04849
Teleplwne: (1) (207) 7633182
Contact. Arlen e Jurewicz
FAX: (1)(207) 2363847
Categories: A8 B4 C4 C6 C7 C8 C9 C10 C11 C12

Third Dimension Arts, LTD
3-D Arts T.In.
1241 Andersen Drive
SuitesC&D
San Raphael, CA 94901
Teleplwne: (1) (415) 485 1730
Contact. Tim Laduca
FAX: (1)(415) 4850435
Categories: A6 AI2 B2 B3 B4 B8

Three-D Light Gallery
107 The Commons
Ithaca, NY
14850
Teleplwne: (1) (607) 2731187
Contact: Eve Walter
Categories: Bl B2 B5

UK Gold Purchasers Inc.DBA Holograms Unltd
7907 Northwest 53rd Street-Suite 193
Miami, FL 33166
Teleplwne: (1) (800) 7227590
Contact: Any/Martin Uram
Categories: BI, B2, B3, B4, B5, B7, B8

Wavefront Reconstruction Compa
6831 Bianca Ave.
Van Nuys,CA
91406
Te!eplume. (1) (818) 3431276
Categories: B7

Wave Mechanics
1533 North Ashland Ave.
Second Floor
Chicago,IL
60622
Te!eplume. (1)(312) 3844860
Categories: B7

Willamette Science & Technology
2300 Centennial Boulevard
P.O. Box 1518
Eugene,OR
97440
Te!eplume. (1)(503) 484 9027
Categories: B4

WonderIight Gallery
2018 R Street, N.W.
Washington, D.C.
20000
Teleplwne: (1) (202) 6676322
Contact: Laurent Bussaut
FAX: (1)(202) 861 0621
Categories: B1 B2 B3 B5

WEST GERMANY

AKS Holographie-Galeria GmbH
Mathildenstrasse 16
4300 Essen 1
Federal Republic of Germany
Telephone. (49) (201) 787622
Categories: B 1 B2

BUYING & SELLING

B1=Gallery
B2=Retail
B3=Fine Art Ltd Editions

B4=Mailorder
B5=Fine Art Holograms
B6=Custom Packaging

B7=Marketing Consultant
B8=Point of Purchase Displays
B9=Custom Lighting

HOL 3 Galerie fur Holographie
Kurfurstendamrn 103
1000 Berlin 31
Federal Republic of Germany
Categories: B1 B2

HOLOGmbH
Holografielabor Osnabruck
MindernerStr.205
04500 Osnabrock
Federal Republic of Germany
Telephone: (0) (541) 7102173
Contact Vito Orazem, T.Luck
FAX: (0) (541) 7102176
Categories: Al A2 A3 A4 A 7 A8 A9 A10 B2 B3 B4
B5 B7 B9 C3 C4 c7 ClO

Holographie Galerie
Amalien Passage 89
Munchen 40 8000
Federal Republic of Germany
Telephone: (49) (89) 285 838
Contact: H. Mielke, E. Fromm
Categories: B1 B2

Holographie Labor
Georgenstrasse 61
0-8000 Munich
Federal Republic of Germany
Telephone: (49) (89) 271 2989
Contact: Mielke
FAX: (49) (89) 271 1375
Categories: Al A2 A3 A5 A7 A8 A10 A12 B2 B5

Museum fur Holographie
& Neue Visuelle Medien
Pletschmiihelenweg 7
5024 Pulheim/Koln
Federal Republic of Germany
Categories: B1 B2

Late Entry:
Holography Institute
PO Box 446
Petaluma, CA 94953-0446
USA
Telephone: (1)(707) 778 1497
Contact: P. Pink
Categories: Al A3 A6 B7. C6 C7 ClO. C12

CHAPTER
6

HOLOGRAPHY IN EDUCATION

Following this brief introduction to holography in education, you will find a chart of those people who are involved with holography in various areas of education. Like the two preceding chapters, the chart lists the individuals and companies and checks which education categories they fit into. In the names and addresses section, all categories in the major areas that a company checks off are listed-i.e. from the Producer, Buying & Selling, Education, and Equipment and Supplies areas.

The role of holography in education

The most accepted area for holography in schools is in the physics class. Most major universities have holography as a course or part of a course. This is something that has not been taken full advantage of by suppliers. There is a need for good all-in-one kits that provide everything including examples of different types of finished holograms. We hope to see this kind of kit in the future.

Rather than being treated as strictly one semester physics course, holography should also be seen as a career path in business, technological development, advanced scientific research and as a new conceptual and material resource for the visual arts and design.

Holography education as a marketing tool

For the producer of holograms, it is financially important that everyone along the line, all the way to the customer, receives some instruction about your hologram.

Suppose you have the good vision to use holography on a product that your research confirmed would be an excellent application. When considering packaging, there are some important concerns for holography that are not concerns for other items in product design. For example, if you do not consider the lighting and display of your hologram product you have diminished the eye catching appeal of your product. You will have created a potential problem that could easily have been prevented by educating your designers. The same holds true for those all the way to the consumer.

As holography becomes better understood, education becomes less of a pressing issue. Now, however, it is most important to set up a simple, quick and effective educating program that supports the use and presentation of your hologram product in shops. Shop managers that know how to present the hologram product well can sell that product better and encourage more hologram products to fly through inventory. The education and support does not have to be elaborate or time-consuming but does have to be present and accurate.

Holography in the classroom

One might ask, "Why is holography not taught more in schools?" Aside from there being too few teachers who realize its range of educational applications, it really is a matter of state and national educational administrators being uninformed of this necessity. Teacher training programs do not begin with the teachers-they begin with the administrators. A coherent, broad-based program for guiding holography into the classroom has to be initiated at the highest levels to make sure all of the other conditions for effectiveness, growth and continuation of programs remain in place permanently.

How will the key decision-making people in schools and businesses understand the necessity of holography, the ease of application, and the rewards of utilizing this medium correctly? There are three steps to making the best presentation for understanding this medium. The first is to expose people to the hologram; the second is to provide simple, accurate information and support ; and the third is to offer a sales presentation of this information with holograms present.

EXPOSURE

To comprehend what holography actually is, it is imperative to see some holograms. It is very possible to meet intelligent people who have heard of holography, have seen photographs of holograms and have read some of the most clearly written descriptions, but in fact do not have any idea of what a hologram is or what it does.

INFORMATION SUPPORT

Immediately beside a hologram on display should be some well written literature designed for the slightly above average and inquiring mind. When preparing your literature, note that holography is not "like" any other image we have ever seen. It is very easy to use the analogy of photography. Using photographs to describe holography is only useful when you are sure the viewers already have a substantial basic understanding of photography. Holography is not like stereoscopic or three-dimensional photography. Even if it was, not much of the general public is familiar enough with three-dimensional photography to understand the basic phenomenon of three-dimensional perception. Analogies should be handled with care.

PRESENTATION

To be certain holography will be understood by most people the third element of presentation has to be present. It is possible to lecture for days and weeks and still leave the audience ignorant of holography. Having an informed salesperson or spokesperson available and presenting the hologram and process, the medium can be comprehended within minutes. The presenter also has the advantage of catching questioning looks from the audience every step of the way and of clarifying steps in the process. The audience can participate and ask questions at the moment of uncertainty. Finally, if you can get the interested customer to a lab classroom, the act of making a hologram is more explanatory than anything else. Perhaps it reinforces the "acceptance" factor in that seeing is believing.

Everyone has been conditioned to see images exist on a flat surface. A three- dimensional picture is still flat but has representational clues to three dimensions; like shading shapes to look like forms or using visual clues to indicate depth-like the diagonal thrusts of perspective. When people are first introduced to a hologram, the brain has some significant reappraisals of what a three-dimensional image and visual reality really are.

HOLOGRAPHY EDUCATION

AUSTRALIA – NETHERLANDS

	C1 MUSEUM	C2 CURATOR	C3 TEACHER TRAINING	C4 HANDS-ON WORKSHOP	C5 COLLEGE CREDIT COURSES	C6 CURRICULUM DEVELOPMENT	C7 PUBLIC SCHOOL PROGRAMS	C8 EDUCATIONAL SUPPORT MATERIALS	C9 COLLEGE LEVEL INSTRUCTOR	C10 TRAVELLING	C11 OUTREACH PROGRAMS	C12 INDEP. EDUC FACILITY
AUSTRALIA												
New Dimension Holographics												
Sydney College of the Arts					✓		✓					
BRAZIL												
Museu Da Imagem E Do Som	✓											
BULGARIA												
Bulgarian Academy of Sciences												
CANADA												
Fringe Research Holographics					✓							
Laval University (Quebec City)			✓	✓	✓			✓		✓		✓
Laval University (Ste Foy)				✓	✓							
Light Construction, Inc.				✓	✓							✓
Ontario College of Art					✓							
Ontario Science Centre					✓							
York University			✓	✓	✓	✓	✓	✓	✓			
DENMARK												
The Holographic Laboratory A/S	✓											
Museum fur Holographie...												
Technical University	✓											
FRANCE												
Holo-Laser	✓						✓			✓	✓	
Institut Holographique de Paris				✓	✓							
Musee De L'Holographie	✓										✓	
ITALY												
Instituto Di Fisica					✓	✓						
Newport: DB Electronic Instrument		✓	✓	✓								
SOI - Society Olografica Italia										✓		
JAPAN												
HoloARP			✓		✓		✓	✓	✓	✓		
Yumiko Shiozaki			✓						✓	✓		
MEXICO												
Hologramas De Mexico										✓		
NETHERLANDS												
Optische Fenomenen								✓				

HOLOGRAPHY EDUCATION

	C1 MUSEUM	C2 CURATOR	C3 TEACHER TRAINING	C4 HANDS-ON WORKSHOP	C5 COLLEGE CREDIT COURSES	C6 CURRICULUM DEVELOPMENT	C7 PUBLIC SCHOOL PROGRAMS	C8 EDUCATIONAL SUPPORT MATERIALS	C9 COLLEGE LEVEL INSTRUCTOR	C10 TRAVELLING	C11 OUTREACH PROGRAMS	C12 INDEP. EDUC FACILITY
NETHERLANDS (continued) –												
Que Sera Sera												
PEOPLES REPUBLIC OF CHINA												
Holography Laboratory				✓								
SPAIN												
University of Alicante					✓							
SWEDEN												
Institute of Physics, LTH					✓							
Royal Institute of Technology					✓							
UNITED KINGDOM												
Ascot Laser Picture Studio				✓					✓			✓
Margaret Benyon				✓								
Darkroom Eight Ltd.				✓								
Holographics International					✓			✓				
Holography Group				✓	✓							
Holography Workshop				✓	✓	✓		✓				
Imperial College of Science				✓	✓							
Interchange Studios												
Clive J. Kocher, FRPS						✓						
Light Engineering												
Loughborough University of Technology					✓							
National Physical Laboratory								✓				
Optical Engineering Group			✓	✓		✓	✓	✓	✓			
Andrew Pepper				✓					✓			
Perception Holography									✓			
David Pizzanelli									✓			
Royal College of Art						✓			✓		✓	✓
Graham Saxby												
Wenyon & Gamble												
UNITED STATES OF AMERICA												
Anait Studio				✓		✓		✓				
Artigliography			✓		✓	✓					✓	✓
Art Institute/Chicago										✓		
Art, Science & Technology Institute	✓											
Laurent Bussaut		✓										

HOLOGRAPHY EDUCATION

UNITED STATES OF AMERICA (continued)	C1 MUSEUM	C2 CURATOR	C3 TEACHER TRAINING	C4 HANDS-ON WORKSHOP	C5 COLLEGE CREDIT COURSES	C6 CURRICULUM DEVELOPMENT	C7 PUBLIC SCHOOL PROGRAMS	C8 EDUCATIONAL SUPPORT MATERIALS	C9 COLLEGE LEVEL INSTRUCTOR	C10 TRAVELLING	C11 OUTREACH PROGRAMS	C12 INDEP. EDUC. FACILITY
Casdin-Silver Holography (Boston)		✓			✓	✓	✓	✓		✓	✓	✓
Casdin-Silver Holog. (Brookline)		✓			✓	✓	✓	✓		✓	✓	✓
Center For Advanced Visual Studies					✓					✓		
Central Michigan University					✓					✓		
Cherry Optical Company	✓											
Chicago Museum of Holography		✓	✓		✓	✓				✓	✓	✓
Clark University									✓			
Davis Publications												
Edward Dietrich		✓	✓		✓	✓						
Edmund Scientific Co.								✓				
F.A.S.T. Elec. Bulletin Board							✓	✓				
Gary Caccione							✓	✓				
Holograny	✓	✓	✓	✓			✓	✓				
Holographic Design, Inc.												
Holographic Visions	✓	✓	✓		✓	✓		✓		✓		✓
Holography News												
Holography Shop of Milwaukee	✓			✓		✓	✓	✓	✓	✓	✓	✓
Holography Workshops		✓		✓		✓						
Integraf												
John Kaufman Photography			✓	✓		✓			✓	✓		
Laser Affiliates		✓		✓		✓						
Laser Arts Society for										✓		
L.A.S.E.R. Co.								✓				✓
Laser Dreams	✓			✓	✓							✓
Laser Institute Of America			✓									✓
Laser Light Designs												✓
Linda Law Holographics		✓	✓	✓				✓		✓	✓	✓
Leonardo (journal)								✓				
Light Impressions Inc.								✓				✓
Los Angeles School of Holography			✓						✓✓			✓
Gerald Marks Studio						✓✓				✓		✓
Massachusetts Institute of Technology					✓	✓		✓				
McCain Marketing												
Media Interface, Ltd.			✓			✓						✓

HOLOGRAPHY EDUCATION

	C1 MUSEUM	C2 CURATOR	C3 TEACHER TRAINING	C4 HANDS-ON WORKSHOP	C5 COLLEGE CREDIT COURSES	C6 CURRICULUM DEVELOPMENT	C7 PUBLIC SCHOOL PROGRAMS	C8 EDUCATIONAL SUPPORT MATERIALS	C9 COLLEGE LEVEL INSTRUCTOR	C10 TRAVELLING	C11 OUTREACH PROGRAMS	C12 INDEP. EDUC FACILITY
U.S.A. (cont'd)—												
Museum Of Holography	✓									✓	✓	
New York Academy of Sciences			✓	✓								✓
New York Hall Of Science	✓											
New York Holographic Laboratory							✓					
New York State Education Department							✓					
Odhner Holographics			✓	✓		✓		✓	✓			✓
Oregon Graduate Center												
Reconnaissance												✓
Rowland Institute for Sciences	✓	✓	✓		✓	✓	✓	✓	✓	✓	✓	✓
School of Holography/Chicago					✓							✓
School of Holography/S Francisco					✓		✓			✓	✓	✓
Robert Sherwood Holog. Design					✓							
Society for Photo-Optical Engineering					✓							
Stanford University Dept. Mech. En.					✓				✓			
Synchronicity Holograms			✓	✓		✓		✓	✓	✓	✓	✓
University of Michigan Dept. Mech En.					✓	✓			✓			
Doris Vila					✓	✓			✓			
Vincennes University				✓	✓	✓	✓	✓	✓	✓	✓	✓
Ed Wesly	✓	✓	✓	✓		✓			✓			
WEST GERMANY												
Dr. Peter Zec										✓		✓
LATE ENTRY: USA												
Holography Institute					✓	✓	✓			✓		✓
LATE ENTRY: SWEDEN												
Holo Media AB/Hologram Museum	✓							✓		✓		

HOLOGRAPHY EDUCATION

C1=Museum
C2=Curator
C3=Teacher Training
C4=Hands-On Workshop

C5=College Credit Courses
C6=Curriculum Development
C7=Public School Programs
C8=Educational Support Materials

C9=College Level Instructor
C10=Traveling Exhibits
C11=Outreach Programs
C12=Independent Educational Facility

AUSTRALIA

New Dimension Holographics
5~72 Pier One
Hickson Road
Sydney, New South Wales 2000
Australia
Telephone: (51) (02) 276063
Contact. Tony Butteriss
Categories: Bl B2 B3 B5 B6 B8 C7

Sydney College of the Arts
Dept of Holography
58 Allen Street
Glebe, Sydney
Australia
Categories: C5

BRAZIL

Museu Da Imagem E Do Som
Ali Europa 158
Sao Paulo 01449
Brazil
Telephone: (55) (11) 8531498
Categories: C1

BULGARIA

Bulgarian Academy of Sciences
Central Lab of Optical Researc
Acad G. Bontchev Street
Sofia 1113 Bulgaria
Telephone: (359) (2) 710 018
Categories: C5

CANADA

Fringe Research Holographics
1179A King Street, West: Suite 008
Toronto, Ontario
Canada M6K 3C5
Telephone: (416) 535 2323
Contact: Michael Sowdon, Dir.
Categories: A2 A3 A7 A8 A11 B1 B5 C4 C8 C10

Laval University
1145 Ave. Des Laurentides
Quebec City, Quebec
Canada GIS 3C2
Telephone: (1) (418) 6872
Contact. Marie Cossette
Categories: Al A2 A3 B5 C3 C4 C5 C12

Laval University
Department of Physics
Pavillion Alexandre Vachon
Sainte F oy, Quebec
CanadaGIK 7P4
Telephone: (1)(418) 6563436
Contact. Prof. RA Lessard
FAX: (1)(418) 65659
Categories: C4C5

Light Construction, Inc.
2154 Dundas Street West
Suite #503
Toronto, Ontario
Canada M6R lX3
Telephone: (1)(416) 5334692
FAX: (1)(416) 5330572
*Categories:*Al A2 A7 A8 AIO B5 C12

Ontario College of Art
100 McCaul Street
Toronto, Ontario
Canada M5T lWI
Telephone: (1) (416) 9775311
Categories: C5

Ontario Science Centre
770 Don Mills Road
Don Mills, Ontario
Canada M3C IT3
Telephone: (1) (416) 429 4100
Categories:. C5

HOLOGRAPHY EDUCATION

C1=Museum	**C5**=College Credit Courses	**C9**=College Level Instructor
C2=Curator	**C6**=Curriculum Development	**C10**=Traveling Exhibits
C3=Teacher Training	**C7**=Public School Programs	**C11**=Outreach Programs
C4=Hands-On Workshop	**C8**=Educational Support Materials	**C12**=Independent Educational Facility

York University
Department of Physics
4700 Keele St
North York, Ontario
Canada M3J IP3
Telephone: (1) (416) 736 2100
Contact Dr. S.B. Joshi
Categories: C3 C4 C5 C6 C7 C8 C9

DENMARK

The Holographic Laboratory A/S
The Danish Technical High School
Building 302
2800 Lyngby
Denmark
Categories: C4

Museum fur Holographie & Neue Visuelle Medien
Valkendorfsgrade 13.1
1151 Copehagen K.
Denmark
Telephone: (45)(1) 93 19 91
Categories: CI

The Technical University of Denmark
Dr Erik Dalsgaard
Physics Laboratory 1, Building
DK 2800 Lyngby
Denmark
Telephnne. (45) (1) 81611
Categories: C5

FRANCE

Holo-Laser
12, Rue De Vouille
75015 Paris, France
Telephone: (33) (1) 45315275
Contact. Dr. J.L. Tribillon
Categories: Al A2 A3 A4 A6 A7 A8 AIO CI CIO Cll

Institut Holographique de Paris
25 Rue Cavendish
75019 Paris
France
Telephone: (33) (1) 47537104
Contact: Marielle Dewatre
Categories: C4

Musee De L'Holographie
Forum des Halles 15 A 21
Grand Balcon - 75001 Paris
France
Telephone: (33)(14) 296 9683
Categories: B1 B2 C1

ITALY

Instituto Di Fisica
Universita Di Roma
P. Ie A. Moro 2
OOI85Roma
Italy
Telephone: (39) (06) 4976287
Categories: C5

Newport: DB Electronic Instrument
Via Torino 5
20032 Cormano
Milano
Italy
TekpIume (39) (92) 32313
Categories: C2 C3 C4 C11

SOl - Society Olografica ltalia
Via degli Eugenii 23
OOI78Roma
Italy
Telephone: (39) (6) 799 0452
Categories: C10

HOLOGRAPHY EDUCATION

C1=Museum	**C5**=College Credit Courses	**C9**=College Level Instructor
C2=Curator	**C6**=Curriculum Development	**C10**=Traveling Exhibits
C3=Teacher Training	**C7**=Public School Programs	**C11**=Outreach Programs
C4=Hands-On Workshop	**C8**=Educational Support Materials	**C12**=Independent Educational Facility

JAPAN

HoloARP
Kamakura Inc.
7-10-8 Ginza Chuo-Ku
Tokyo
Japan
Telephone:. (81) (03) 5748307
Contact: Yumiko Shiozaki
FAX: (81) (03) 5748377
Categories: Al A2 A3 A4 A6 Bl B3 B5 C3 C6 C9

Yumiko Shiozaki
7-5-18-Ryoke Urawa
Saitama
Japan
Telephone: (81) (0488) 31 7723
Contact. Yumiko Shiozaki
Categories: A2 B5 C3 C5 C7 C8 C9 ClO

MEXICO

Hologramas De Mexico
PINO 343, Local 3
COL. STA. MA LA RIBERA
06400 Mexico, D.F.
Mexico
Telephone: (52) (905) 5479046
Contact. Dan Lieberman
FAX (52) (905) 5474084
Categories: Al A2 A4 A5 A6 A8 A9 A12 B2 B3 B6 C10

NETHERLANDS

Optische Fenomenen
Nederlandse Stichting Voor
Waarneming & Holografie
Warenarburg 44
NL 2907 CL Capelle a/ d Ijssel,
Contact: Jan Broeders
Categories: C8

Que Sera Sera
P.O. Box 29
9700 AA Groningen
The Netherlands
Telephone:. (31)(050) 140417
Contact:. H.T. Vogd
FAX: (31)(050) 144142
Categories: A12 B3 C4 D3 D4

PEOPLES REPUBLIC OF CHINA

Holography Laboratory
Department of Applied Physics
Beijing Inst. of Posts & Tech.
Beijing
People's Republic of China
Telephone: (86) (1) 668 1255
Contact. Da-Hsiung Hsu
Categories: C5

SPAIN

University of Alicante
Departmen t of Applied Physics
Facultad de Ciencias
Alicante Apdo 99
Spain
Telephone: (34) (65) 568 1150
Categories: C5

SWEDEN

Institute of Physics, LTH
Box 118
22100 Lund
Sweden
Telephone: (46) (46) 107656
Categories: C5 D5

HOLOGRAPHY EDUCATION

C1=Museum	**C5**=College Credit Courses	**C9**=College Level Instructor
C2=Curator	**C6**=Curriculum Development	**C10**=Traveling Exhibits
C3=Teacher Training	**C7**=Public School Programs	**C11**=Outreach Programs
C4=Hands-On Workshop	**C8**=Educational Support Materials	**C12**=Independent Educational Facility

Royal Institute of Technology
Department of Industrial Meter
S-10044 Stockholm
Sweden
Telephone: (46) (8) 7879175
Categories: C5

UNITED KINGDOM

Ascot Laser Picture Studio
27 Upper Village Road
Sunninghill, Ascot
Berkshire S1..5 7Aj
England, United Kingdom
Telephone: (44) (0990) 21789
Contact: Mr. Brode!
Categories: A2 A3 A7 A8 C4 C9 C12

Margaret Benyon
Holography Studio
40 Springdale Avenue
Broadstone, Dorset
BH18 9EU, England
Telephone: (0202) 698067
Categories: A2 A3 A 7 A8 B5 C4

Darkroom Eight Ltd.
Unit 8 - Impress house
Vale Grove, Acton
London W3 7QH
United Kingdom
Telephone:(44) (1) 7492218
Categories: Al C4

Holographies International
BCM-Holographics
London WCIN 3XX
England
United Kingdom
Telephone: (44) (1) 584 4508
Categories: C8

Holography Group
University of Oxford
Dept. of Engineering Science
Parks Road, Oxford
OXI 3Pj, England
Telephone: (44) (865) 273805
Categories: C5

Holography Workshop
Goldsmith College
Millard Bldg., Cormont Road
London SE5 9RG
England, United Kingdom
Telephone: (44) (1) 7333216
Categories: C4

Imperial College of Science
Optics Section
Blackett Laboratory
London SW72BZ
England, United Kingdom
Telephone:. (44) (1) 589 5111
Categories: C5

Interchange Studios
15 Wilkin Street
London NW5 3NG
England, United Kingdom
Telephone: (44)(1) 2679421
Categories: C4

Clive J. Kocher, FRPS
Dept of Photography
Salisbury College of Art
Southampton Road, Salisbury
Wiltshire, England, U.K.
Telephone: (44) (722) 23711
Categories: C8

HOLOGRAPHY EDUCATION

C1=Museum	**C5**=College Credit Courses	**C9**=College Level Instructor
C2=Curator	**C6**=Curriculum Development	**C10**=Traveling Exhibits
C3=Teacher Training	**C7**=Public School Programs	**C11**=Outreach Programs
C4=Hands-On Workshop	**C8**=Educational Support Materials	**C12**=Independent Educational Facility

Light Engineering
12 New StJohns
St. Helier,Jersey
Channel Islands England
Telephone: (44) (534) 21758
Categories: CA

Loughborough University of Technology
Dept. of Mechanical Engineerin
Loughborough, Leicester
England LEll 3TU
Telephone: (44)(509) 263 171
Categories: C5

National Physical Laboratory
Teddington, Middlesex
England TWII OLW
United Kingdom
Telephone:: (44) (1) 9773222
Categories: C8

Optical Engineering Group
Dept. of Mechanical Engineerin
Loughborough Univ. of Tech nolo
Loughborough, Leicestershire,
England, United Kingdom
Telephone: (44) (0509) 223222
Contact: John Tyror
FAX: (44) (0509) 231 983
Categories: Al A2 A4 A5 A6 A 7 A8 A9 A11 AI2 C3
C4 C8

Andrew Pepper
22 Haldane Road
London E6 3.IJ
England
United Kingdom
Telephone: (44) (1) 4711609
FAX: *(44)(1) 3181439*
Categories: A2 A3 A8 C6 C7 C9

Perception Holography
Thornton Marketing Ltd.
Aketon Close, Haggs Lane
Follifoot, Harrogate
Facility
North Yorks.,England HG3 IA2
Telephone: (44) (93) 782323
Contact Mike Burridge
Categories: Al A 7 A8 A11 CA C9

David Pizzanelli
4 Maculay Road
London, SW4 OQX
England
United Kingdom
Telephone: (44)(1) 6271140
Categories: Al A2 A3 A4 A6 A7 A8 A9 B7 C9

Royal College of Art
Holography Unit
Kensington Gore
London SW7 2EU
England
Telephone: (44)(1) 5845020
Categories: C11 C12

Graham Saxby
3 Honor Avenue
Goldthorn Park, Wolverhampton
West Midlands, WV45HF
England, United Kingdom
Telephone: (44) (902) 341 291
Categories: C9

Wenyon & Gamble
8 Berry Street
London ECIV OAU
England
United Kingdom
Telephone: (44)(1) 2511797
Categories: A2 C6

HOLOGRAPHY EDUCATION

C1=Museum	**C5**=College Credit Courses	**C9**=College Level Instructor
C2=Curator	**C6**=Curriculum Development	**C10**=Traveling Exhibits
C3=Teacher Training	**C7**=Public School Programs	**C11**=Outreach Programs
C4=Hands-On Workshop	**C8**=Educational Support Materials	**C12**=Independent Educational Facility

UNITED STATES OF AMERICA

Anait Studio
1685 Fernald Point Lane
Santa Barbara, Calif.
93108
Telephone: (1)(805) 9695666
Contact. Anait
*Categories:*A2 A3 All B5 C4 C12

Artigliography
7130 Mohawk West Drive
Indianapolis, IN
46236
USA
Telephone: (1) (317) 8230069
Contact: Kerry S. Brown
Categories: A2 A7 A8 B3 B5 C3 C6 C8 Cll

Art Institute of Chicago School
Holography Department
Columbus and Jackson Boulevard
Chicago, IL 60603
Telephone: (1) (312) 3847347
Cantact. Dept Head
Categories: C5

Art, Science & Technology Institute
2018 R Street N.W.
Washington D.C. 20009
USA
Telephone: (1) (202) 6676322
FAX: (1)(202) 861 0621
Categories: C1 C10

Laurent Bussaut
Research Director, Curator
2018 R Street, N.W.
Washington, D.C.
20009
Telephone: (1)(202) 6676322
FAX: (1) (202) 861 0621
Categories:. C2

Casdin-Silver Holography
51 Melcher Street
Studio 501
Boston, MA 02210
Facility
Telephone: (1)(617) 4234717
Contact: Harriet Casdin-Silver
Categories: Al A2 A3 A4 AS A7 A8 A9 AIO All B2 B3 B5 B6 B8 B9 C2 C5 C6 C7 C9 CIO CI1

Casdin-Silver Holography
99 Pond Avenue
Suite D403
Brookline, MA
02146
Telephone: (1) (617) 7396869
Contact: Harriet Casdin-Silver
Categories: A1 A2 A3 A4 AS A7 A8 A9 A10 A11 B2 B3 B5 B6 B8 B9 C2 C5 C6 C7 C9 C10 C11

Center For Advanced Visual Studies
Massachusetts Institute of Technology
40 Massachusetts Avenue
Cambridge, MA 02139
Telephone: (1) (617) 2534478
Categories: C5

Central Michigan University
Art Department
Attn: Richard Kline
Mt Pleasant, MI
48859
Telephone: (1) (517) 7743025
Categories: C5

Cherry Optical Company
2047 Blucher Valley Road
Sebastopol, CA
95472
Telephone: (1) (707) 8237171
Contact G.Cherry IN .Gorglione
Categories: A1 A2 A3 A4 A7 A8 A9 AIO B3 B4 B7 C2 C3 C6 C10

HOLOGRAPHY EDUCATION

C1=Museum	**C5**=College Credit Courses	**C9**=College Level Instructor
C2=Curator	**C6**=Curriculum Development	**C10**=Traveling Exhibits
C3=Teacher Training	**C7**=Public School Programs	**C11**=Outreach Programs
C4=Hands-On Workshop	**C8**=Educational Support Materials	**C12**=Independent Educational Facility

Chicago Museum of Holography
1134 West Washington Boulevard
Chicago,IL
60607
Telephone: (1) (312) 2261007
Cantact:. L. Billings
Categories: Bl B2 B3 B5 Cl C2 C7 CIO C11 C12

Clark University
Department of Physics
Worcester, MA 01610
Telephone: (1)(617) 7937756
Categories: C5

Davis Publications
Science & Mechanics Division
380 Lexington Avenue
New York, NY 10017
Telephone: (1)(212) 5579100
Categories: C8

Edward Dietrich
2036 West Haddon
Chicago,IL
60622
Telephone: (1) (312) 2920770
Contact. Edward Dietrich
Categories: A2 C 3C5 C6 C8 C9

Edmund Scientific Co.
101 East Gloucester Pike
Barrington, r.rr 08007
Telephone: (1) (609) 547 8900
Categories: B2 C8

F.A.S.T. Elec. Bulletin Board
P.O. Box 421704
San Francisco, CA 94142-1704
Telephone:. (1) (415) 8458306
Contact: Elizabeth Crumley
FAX: (1)(415) 8416311
Categories: C8

Gary Gaccione
2680 Lee Place
Belmore, NY 10710
Categories: C7

Hologramy
116 St Marks PI. #3
New York, NY 10009
Telephone: (1)(212) 529 3955
Contact: S. Lloyd Facility
Categories: C2 C3 C4 C6 c7 C8 Cll

Holographic Design, Inc.
1084 North Delaware Avenue
Philadelphia, PA 19125
Telephone: (1) (215) 4259220
Conlact: D.Miller
Categories: Al A4 A6 A12 B3 B6 B7 B8 C7

Holographic Visions
300 South Grand Avenue
Los Angeles, CA
90071
Telephone: (1) (213) 6877171
Contact: Mark Merrill
Categories: B1 B2 B3 B4 B5 B6 B7 B8 B9 C1 C2 C8 C11

Holography News
3932 Mckinley Street, N.W.
Washington, D.C. 20015
Contact:. L Kontnick
Categories: Al B7 C12

Holography Shop of Milwaukee
5644 Parking Street
Greendale, WI
53129
USA
Telephone:. (1)(414) 4216767
Categories: C8

Holography solves NDT problems *nothing else can solve*

...and Newport makes it push-button easy

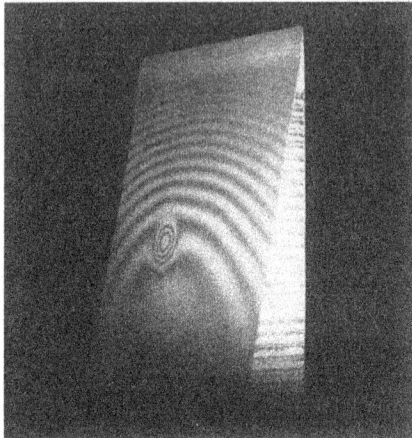

Debonds and Composite flaws are determined in both size and location with holography's ability to test the entire part in a single view. Holography's high sensitivity allows detection of flaws as small as an individual honeycomb cell.

Thermal Distortions can be observed throughout the entire deformation cycle. You will see more than just where it ended up, you will also see how it got there.

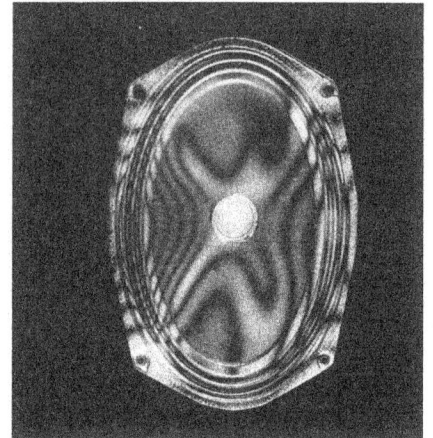

Modal Analysis is simplified since you can actually see the shape of resonant modes as they occur. No more guessing about mode frequency, amplitude, or shape, and no more mechanical probes to perturb modal characteristics.

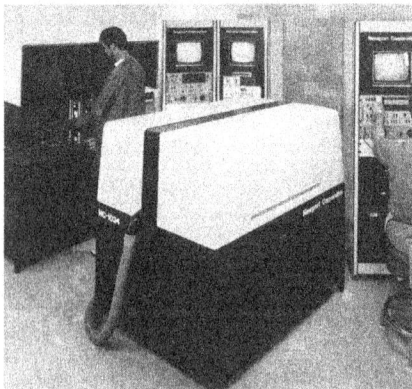

The HC-1034 Holographic System in operation for quality control testing of audio speakers. This and other systems are featured in our new catalog. Call for yours today.

We stock Agfa and Kodak Holographic Film

714/965-5416
Newport Corporation
18235 Mt. Baldy Circle
Fountain Valley, CA 92708
Europe: Newport GmbH, Ph. 06151-26116
U.K.: Newport Ltd., Ph. 05827-69995

Newport

HOLOGRAPHY EDUCATION

C1=Museum	**C5**=College Credit Courses	**C9**=College Level Instructor
C2=Curator	**C6**=Curriculum Development	**C10**=Traveling Exhibits
C3=Teacher Training	**C7**=Public School Programs	**C11**=Outreach Programs
C4=Hands-On Workshop	**C8**=Educational Support Materials	**C12**=Independent Educational Facility

Holography Workshops
Lake Forest College
Sheridan and Collge Road
Lake Forest. IL
00045
Telephone: (1) (312) 2343100
Contact:. Tung H. Jeong
Categories:: A1 A2 A4 A7 A8 A9 A10 A11 A12 B1 C1
C2 C3 C4 C5 C6 c7 C8 C9 ClO C11 C12

Integraf
745 North Waukegan Road
Lake Forest. IL
00045
Telephone:. (1)(312) 2343756
Contact: Anna Wong
Categories: B2 B3 B4 B7 C3 C6 C7 DI D3 D6

John Kaufman Photography
P.O. Box 477
Point Reyes Station, CA 94956
Telephone: (1) (415) 6631216
Categories: A2 A3 A8 AlO B5 C4

Laser Affiliates
2047 Blucher Valley Road
Sebastopol, CA
95472
Telephone: (1)(707) 8237171
Contact: Nancy Gorglione
Categories: A2 A3 A4 A7 A8 A9 AIO C2 C3 C6 C9
C10

Laser Arts Society for Education & Research
P.O. Box 42083
San Francisco, Calif.
94101
Contact:. Brad Can tos
Categories: C8

LAS.E.R. Co.
1900 Gore Drive
Haymarket, VA
22000
Telephone: (1)(703) 7542526
Contact Jim Bowman
Categories: A2 A3 A7 A8 A9 B2 B5 B9 C4 C12

Laser Dreams
Attn: Nancy J. Gorglione
P.O. Box 326
Forestville, CA
95436
Telephone: (1) (707) 8236104
Categories: A2 A3 A4 A7 A8 A9 AID B2 B3 B6 B9
C1

Laser Institute Of America
Education Director
5151 Monroe Street-Suite 102W
Toledo,OH 43623
Telephone: (1)(419) 8828706
Categories: C12 D2

Laser Light Designs
2412 Kennedy Way
Antioch,CA
94500
Telephone: (1)(415) 7543144
Contact: Michael Malott
Categories: Bl B2 B3 B4 B7 B8 ClO

Linda Law Holographies
8 Crescent Drive
Huntington, NY
11743
USA
Telephone: (1) (516) 351 6056
Contact:. linda Law
Categories: A2 A3 A 7 A8 A9 C2 C3 C4 C6 C7 C8
C11 C12

Leonardo (Journal)
P.O. Box 75
1442A Walnut Street
Berkeley, CA 94709
Telephone: (1) (415) 8458306
Contact: Dr. R. Malina
FAX:. (1)(415) 8416311
Categories: C8

MUSEUM OF HOLOGRAPHY–CHICAGO

TWELVE YEARS

OF

HOLOGRAPHIC LEADERSHIP

AWARD WINNING INSTITUTION

MUSEUM – SCHOOL

SCHOOL OF HOLOGRAPHY

COLLEGE CREDIT COURSES

WORKSHOPS – TUTORIALS

INTENSIVE STUDY COURSES

1134 W. WASHINGTON BLVD.
CHICAGO, IL 60607 (312) 226-1007

CHICAGO

MUSEUM
OF
HOLOGRAPHY

HOLOGRAPHY EDUCATION

C1=Museum	**C5**=College Credit Courses	**C9**=College Level Instructor
C2=Curator	**C6**=Curriculum Development	**C10**=Traveling Exhibits
C3=Teacher Training	**C7**=Public School Programs	**C11**=Outreach Programs
C4=Hands-On Workshop	**C8**=Educational Support Materials	**C12**=Independent Educational Facility

Light Impressions Inc.
149 B Josephine Street
Santa Cruz, CA
95000
Telephone: (1) (408) 458 1991
Contact: Kathryn S. Long, Mgr.
FAX: (1)(408) 4583338
Categories: Al AS A6 B2 B3 B4 B6 B7 C8

Los Angeles School of Holography
P.O. Box 851
Woodland Hills, CA 91365
Telephone: (1)(818) 7031111
Contact J. Fox
FAX: (1)(818) 7031182
Categories: C3 C4 ClO C12

Gerald Marks Studio
29 West 26th Street
New York, NY
10010
USA.
Telephone: (1) (212) 8895994
Categories: Al A2 A3 A4 AS C9

Massachusetts Institute of Technology
M.I.T. Media Laboratory
20 Ames Street, EI5-416
Cambridge, MA
02139
Telephone: (1)(617) 2530632
Contact. Jane F. White
FAX: (1) (617) 2586264
Categories: C5 C6 C9

McCain Marketing
10962 North Wauwatosa Road
76W
Mequon, WI
53002
Telephone: (1)(414) 2424023
Contact: Richard McCain
Categories: B4 B7 C6 C8 C12

Media Interface, Ltd.
167 Garfield Place
Brooklyn, NY
11215
Facility
Telephone: (1) (718) 7884012
Contact. Ronald R. Erickson
Categories: Al A4 A9 B9 C3 C6

Museum Of Holography
11 Mercer Street
New York, NY
10013
Telephone: (1) (212) 9250581
Contact: Martha Tomko, Dir.
FAX: (1)(212) 8401663
Categories: C1 C8 C10 C11

New York Academy of Sciences
2 East 63rd Street
New York, NY
10021
USA.
Telephone: (1) (212) 838 0230
Categories: C12

New York Hall Of Science
47-01111th Street
Corona,NY
11368
USA.
Telephone:(1) (718) 699 0005
Categories: Cl C7

New York Holographic Laboratory
P.O.Box 20391
Tomkins Square Station
New York, NY
10000
Telephone: (1)(212) 2549774
Contact: Dan Schweitzer
FAX: (1) (212) 6741007
Categories: Al A2 A3 A6 A7 A8 B1 B5 C3 C4

HOLOGRAPHY EDUCATION

C1=Museum	**C5**=College Credit Courses	**C9**=College Level Instructor
C2=Curator	**C6**=Curriculum Development	**C10**=Traveling Exhibits
C3=Teacher Training	**C7**=Public School Programs	**C11**=Outreach Programs
C4=Hands-On Workshop	**C8**=Educational Support Materials	**C12**=Independent Educational Facility

New York State Department of Education
Room 681
Albany, New York
12234
Telephone:. (1) (518) 4745932
Contact: Bob Reals
Categories: C7

Odhner Holographies
833 Laurel Avenue
Orlando,FL
32803
USA
Telephone:(1) (407) 894 7966
Contact: Jefferson Odhner
Categories: Al A2 A7 A8 C3 C4 C6 C8 C9 C12

Oregon Graduate Center
Attn: Jack Biles
19600 N.W. Walker Road
Beaverton, OR
97006
Telephone: (1)(503) 645 1121
Categories: C8

Reconnaissance
3932 McKinley Street, N.W.
Washington, D.C.
20015
USA
Contact:. L. Kontnick
FAX:. (1) (703) 764-6398
Categories: Al B7 C12

Rowland Institute for Sciences
100 Cambridge Parkway
Cambridge, MA
02142
Telephone: (1)(617) 4974657
Contact. J-M Fournier
Categories: C12

School of Holography/Chicago
1134 West Washington Boulevard
Chicago,IL
60607
Telephone:(1) (312) 2261007
Contact: L. Billings, Dir.
Categories: B1 B2 B3 B4 B5 B7 C1 C2 C3 C4 C5 C6 C7 C8 C9 C10 C11 C12 D1 D6

School of Holography
Attn: Sharon McCormack
550 Shotwell
San Francisco, CA
94110
Telephone: (1)(415) 8243769
Categories: C12

Robert Sherwood Holographic Design
400 West Erie Street
Chicago,IL
00610
USA
Telephone: (1)(312) 944 0784
Contact: K.Kellison,R.Sherwood,
FAX:. (1) (215) 4259221
Categories: Al A4 A6 A12 B6 B7 B8 B9 C7 Cll

Society for Photo-Optical Instrumentation
Engineers
P.O. Box 10
Bellingham, WA
98227
Telephone: (1)(206) 6763290
Categories: C10

Stanford University
Department of Mechanical Engineering
Building 570 Room 571C
Stanford, CA 94301
Telephone: (1) (415) 7233243
Categories: C5

HOLOGRAPHY EDUCATION

C1=Museum
C2=Curator
C3=Teacher Training
C4=Hands-On Workshop

C5=College Credit Courses
C6=Curriculum Development
C7=Public School Programs
C8=Educational Support Materials

C9=College Level Instructor
C10=Traveling Exhibits
C11=Outreach Programs
C12=Independent Educational
Facility

Synchronicity Holograms
Box 4235
Lincolnville, ME 04849
Telephone: (1)(207) 7633182
Contact ArleneJurewicz
FAX: (1)(207) 2363847
Categories: A8 B4 C4 C6 C7 C8 rn ClO Cll C12

The University of Michigan
College of Engineering
Ann Arbor, MI 48109
Telephone:. (1) (313) 764 2390
Contact: Dr. Emmett Leith
Categories: C9 C5

Doris Vila
157 East 33rd Street
New York, NY 10016
Teleplwne. (1)(212) 6865387
Categories: Al A2 A3 A4 A7 A9 AI0

C4C5 C6rn
Vincennes University
1002 North First Street
Vincennes, IN 47591
Telephone: (1)(812) 8855294
Contact: Richard Duesterberg
Categories: C5 C9

EdWesly
5331 North Kenmore Avenue
Chicago, IL 60640
Telephone: (1) (312) 7841669
Categories: A2 A3 A7 A8 All Cl C3
C4 C5 C6 C7 C8 rn CIO Cll C12

WEST GERMANY

Dr. Peter Zec
Lerchenstrasse 142 a
D-4500 Osnabruck
Federal Republic of Germany
Telephone: (49) (0541) 186059
Telex. (49)(0541) 186 059
FAX: (49) (0541) 7102176
Categories: C2 C3 C4 C5 C6 C9 CIO CI2

Late entry:
Holography Institute
PO Box 446
Petaluma, CA 94953-0446 USA
Telephone: (1)(707) 7781497
Contact: P.Pink
Categories: Al A3 A6 B7 C6 C7 C10 C12

Holo Media AB/Hologram Museum
Box 45012
Drottninggatan 100
Stockholm
Sweden
Telephone: (46) (08) 105465
Contact: Mona Forsberg
FAX: (46) (08) 107638
Categories: Al A6 A11 A12 B1 B2 B3 B4 B5 B6 B7 B8 B9
C1 C8 C10

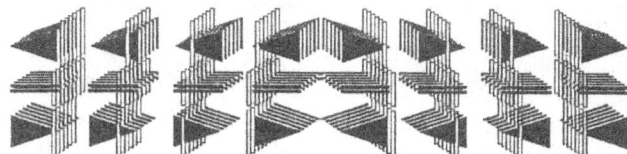

Art, Science and Technology Institute
THE HOLOGRAPHY COLLECTION
Permanent and rotating exhibits
Awards to encourage artistic creativity in holography

• Educational programs
to develop the understanding of holography

• Group tours, seminars, lectures

• Workshops, as well as theoretical and practical sessions
to train new holographers, including a certificate of
completion and a one year of technical assistance.

• Pulsed laser portraiture services
• Traveling exhibits
• Research programs

ASTI
2018 R Street, N.W., Washington, D.C. 20009
Tel. (202) 667-6322 — Fax. 202-861-0621

CHAPTER
7

EQUIPMENT & SUPPLIES

Holography suppliers provide services or materials for holographers. These services and materials include all items used in: hologram laser labs and darkrooms. The chart and names and address listings follow this brief introduction.

It definitely pays to shop around for whatever supplies you may require. Depending on what you want, prices vary widely. On the low end of things, if you are setting up a small holography lab in your basement or at a school it will not be necessary to buy a 50mw Helium Neon laser. A 5mw HeNe laser would suffice. Among the types of 5mw lasers you find the self-contained laser or a separate laser head and power supply. You find more wavelengths than red in HeNe lasers, linear and non-linear polarization, and different "modes". Before you buy anything, ask the advice of a holography consultant or a supply and equipment person in the field. A salesperson for one brand of laser will be happy to tell you the shortcomings
of a competitor's laser.

Ask for advice from qualified hologram producers who have been working in holography for at least three years. You will save a great deal of money and get the proper equipment the first time. In addition you will have good insight into selecting other supplies as you progress. For certain materials or equipment, sometimes the higher price has to be paid but in many circumstances selecting the inexpensive materials actually may better serve your purposes.
Laser Lab
If you are a hobbyist or educator, all of your materials except holography film and plate holders can sometimes
be purchased at one company. If you would like

to make a good hologram inexpensively, buying the components individually after talking to a hologram producer is the surest route to satisfaction.

If you wish to begin a production or research business in holography, again, find a qualified, experienced holographer. It is still commonplace to see expensive equipment not being used correctly, which shows a holographer has not been consulted. If you are truly serious, hire a holography consultant. Not all laser labs are the same and several volumes could be written about all of the useful equipment you could buy. The simplest laser lab equipment (laser included) for the exposure of holograms can be as little as (US) $500.00 to $1000.00. A truly professional lab that is to be used in research or holography production could cost as much as a half million dollars. The equipment you'll need depends upon your goals.

For the laser lab you need a dark, stable environment with low noise and no air circulation during the exposures. In most cases you will need a vibration isolation table that you can build inexpensively or buy expensively. You will need a laser that is a continuous
wave (CW) type for making most holograms or an expensive pulse laser for recording live subjects and making more sophisticated holograms. Lenses should be appropriate for the power and wavelength of laser you are using. The same applies to the front surface mirrors and any other equipment on the table. There are several books on making holograms that suggest equipment for different purposes.

Be sure that everyone in the lab is briefed on laser safety. Most of the information regarding safety for the small lasers is common sense but you should post information on the walls so that everyone is informed. In the more powerful laser labs it is absolutely imperative to have an experienced person operate the laser. You can receive more information by writing to the state and federal departments governing laser use and laser safety.

Holography Darkroom

The materials for the darkroom are much less expensive than the laser and most of the laser lab equipment. Included among darkroom equipment for simple home or school setups is: a chemical scale, processing chemistry, clean white trays and a safelight. Proper ventilation is imperative. Protection for all exposed parts of the body is also imperative. Good rubber gloves, acid-resistant chemical aprons, a filtered mask, a lab sink with a splash protector, chemical-absorbing mats in the mixing area, a time clock with a safe, luminescent face, electricity set

away from the water area, no flammable chemistry, and so on are all priorities of a well-run darkroom, no matter what the size.

Holography darkrooms are no more dangerous than photography darkrooms if the basic rules of safety are heeded. Any substance is potentially dangerous if used incorrectly. Be sure to use the most recently recommended chemistries in holography. Some of the earlier uses of chemicals that contained mercury, bromine, methyl alcohol, xylene, parabenzoquinone (PBQ) should be avoided. There are new, safer chemicals that have been found to actually bring better final hologram results. The Proceedings of the Lake Forest College International Symposia on Holography listed in the bibliography is an excellent resource for chemistry processing techniques and a host of other pertinent issues. You can also contact film manufacturers who would love to supply you with technical information on these matters. Books and magazines are another good source.

USTRALIA — UNITED KINGDOM

	D1 FILM	D2 LASERS	D3 MIRRORS	D4 LENSES	D5 LABWARE	D6 PHOTO-CHEMISTRY	D7 LIGHT METERS	D8 SHUTTERS	D9 POSITIONING EQUIPMENT	D10 SPATIAL FILTERS	D11 PINHOLES	D12 VIBRATION ISOLATION TABLES
AUSTRALIA												
wport-Quentron Optics PTY Ltd		✓										
dio Shack/Tandy Corporation												
BELGIUM												
fa-Gevaert	✓						✓					
CANADA												
nadian Holographic Society												
ling Scientific Ltd.		✓										
DENMARK												
w Dimensions Laser Systems A		✓										
EGYPT												
wport: Scientific & Trading		✓	✓	✓								
FRANCE												
ling S.A.R.L.												
ro Electro-Optics, S.A.		✓	✓	✓								
iversite Louis Pasteur					✓							
ITALY												
herent S.R.L.		✓										
ling Italia		✓										
G Laser Ltd.		✓										
JAPAN												
ji Photo Optical Co., Ltd.			✓	✓								
wport: Kyokuto Boeki Kaisha		✓	✓	✓				✓				
detoshi Katsuma												
NETHERLANDS												
herent, B.V.	✓	✓										
tch Holographic Laboratory			✓	✓								
e Sera Sera												
SWEDEN												
stitute of Physics, LTH				✓	✓							
UNITED KINGDOM												
Electro-Optics Ltd.	✓	✓					✓		✓			
fa-Gevaert Ltd.		✓										
herent, Ltd.												

UNITED KINGDOM (continued) — UNITED STATES OF AMERICA

	D1 FILM	D2 LASERS	D3 MIRRORS	D4 LENSES	D5 LABWARE	D6 PHOTO-CHEMISTRY	D7 LIGHT METERS	D8 SHUTTERS	D9 POSITIONING EQUIPMENT	D10 SPATIAL FILTERS	D11 PINHOLES	D12 VIBRATION ISOLATION TABLES
CSI			✓									
Datasights Ltd.			✓									
Davin Optical Ltd.		✓	✓									
Ealing Electro-Optics.			✓	✓								
Electro Optics Developments Ltd.			✓	✓				✓	✓	✓	✓	✓
Galvoptics Ltd.			✓	✓								
Holofax Limited			✓			✓						✓
Howard Smith Precision Optics			✓									
Ilford Limited	✓											
Kendall Hyde Ltd.												
Laser Instrumentation Ltd.		✓	✓									
Maplin Electronics Supplies Ltd.		✓										
Newport:METAX LTD.		✓										
Radio Shack/Tandy Corporation		✓					✓					
Siemens Ltd.		✓										
Spectra Physics Ltd.		✓	✓									
Vinten Electro Optics Ltd.		✓										
Wentworth Laboratories Ltd.			✓	✓		✓		✓	✓	✓	✓	✓
UNITED STATES OF AMERICA												
Agfa-Gavaert Inc.	✓											
Apollo Lasers Inc.		✓										
City Chemical						✓						
Coherent Inc.		✓										
Continental Optics			✓									
Coulter Optical Company			✓									
CVI Laser Corp.		✓										
Ealing Corporation			✓									
Eastman Kodak	✓											
Electro Optical Industries, Inc.			✓									
G.M. Vacuum Coating Lab, Inc.			✓									
Holographic Film Company	✓											
Holo-Spectra		✓	✓	✓	✓	✓	✓	✓	✓	✓	✓	✓
Hughes Aircraft Co.		✓									✓	
Illinois Valley Magnetic Resonance		✓										

Start Your Next Hologram With A Powerful Advantage.

A surprising thing happens when you put a Coherent Innova® 90 ion laser in your lab.

You get more work done, and it gets done faster.

Because while the Innova 90 costs about the same as competitive lasers, it pumps out 20% more power—6 watts all lines, or a guaranteed minimum 1.2 watts single-frequency at 514.5 nm.

And the high-thermal-mass SuperInvar resonator provides unmatched frequency stability for long coherence lengths.

So you always have ample power and stability—even for your most challenging exposures.

And the Innova 90 matches this extra power with unequalled reliability. Independent studies prove that the tungsten-disk/metal-ceramic technology pioneered by Coherent is superior to other ion technologies such as BeO, and that Coherent ion plasma tubes last at least 25% longer than the closest competitive metal-ceramic plasma tubes.

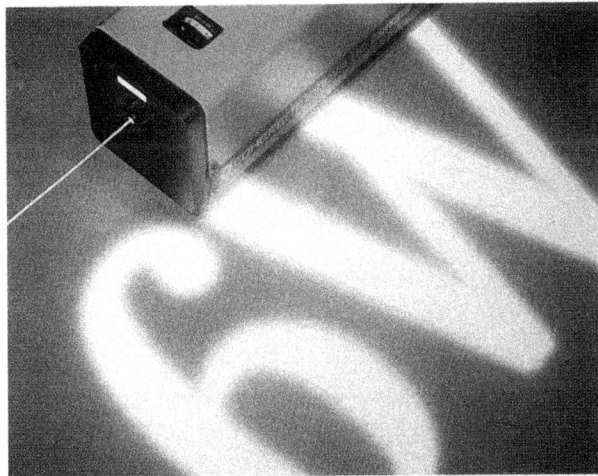

Furthermore, we're the only ion laser manufacturer that guarantees to have replacement plasma tubes in stock at all times, to support our team of field service engineers.

Performance, proven reliability and outstanding customer support guarantee you the highest uptime possible. That's what makes our laser users the most productive people in the industry.

Get more details now: In the United States call 800-527-3786; United Kingdom (0223) 68501; West Germany (06074) 9140; Japan (03) 639-9871. Or write, Coherent Laser Products Division, 3210 Porter Drive, P.O. Box 10321, Palo Alto, CA 94303.

We'll show you the difference power and reliability can make.

COHERENT LASER PRODUCTS

✳®COHERENT®

NITED STATES OF AMERICA (continued) —
EST GERMANY

	D1 FILM	D2 LASERS	D3 MIRRORS	D4 LENSES	D5 LABWARE	D6 PHOTO-CHEMISTRY	D7 LIGHT METERS	D8 SHUTTERS	D9 POSITIONING EQUIPMENT	D10 SPATIAL FILTERS	D11 PINHOLES	D12 VIBRATION ISOLATION TABLES
egraf	✓					✓						
ernational Dikrotek, Inc.	✓		✓									
Laser Technology		✓										
lon Incorporated	✓	✓	✓	✓	✓	✓	✓	✓	✓	✓	✓	✓
ser Optical Systems, Inc.		✓	✓✓	✓✓								
nbda/Ten Optics		✓✓										
ser Institute Of America		✓										
er Resale Inc.												
trologic Instruments, Inc.			✓	✓								
tutoyo Measuring Instruments					✓✓							
wport Research Corporation		✓✓✓	✓✓	✓		✓	✓	✓				
rland Products, Inc.				✓✓								
tics Plus Inc.	✓	✓	✓	✓								
tical Purchasing Directory												
ctro-Optical Specialist						✓					✓	
timation												
otographic Formulary						✓						
S Electro-Optics		✓✓	✓✓✓			✓	✓✓					
aroid Corporation	✓		✓	✓		✓						
lio Shack	✓											
ence & Mechanics Instrument	✓											
pley Chemical Co.												
ctra Lumicon		✓✓										
ctra-Physics		✓✓✓	✓	✓								
ctrogon AB												
WEST GERMANY												
ing GmbH												
wportCarl Baasel Lasertechn												
ctra-Physics GmbH												

EQUIPMENT & SUPPLIES

D1=Film	**D5**=Labware	**D9**=Positioning Equipment
D2=Lasers	**D6**=Photochemistry	**D10**=Spatial Filters
D3=Mirrors	**D7**=Light Meters	**D11**=Pinholes
D4=Lenses	**D8**=Shutters	**D12**=Vibration Isolation Tables

AUSTRALIA

Newport-Quentron Optics PTY Ltd
Allenby Gardens
South Australia 5009
Australia
Telephone: (61) (8) 466121
Categories: D2

Radio Shack/Tandy Corporation
280-316 Victoria Road
Rydalmere
New South Wales
2116 Australia
Categories: D7

BELGIUM

Agfa Gevaert
Holography Film Dept
B251O
Mortsel, Belgium
Telephone: (32) (3) 444 8242
Categories: D1

CANADA

Canadian Holographic Society
3577 Rue de Bullion
Montreal, Quebec
Canada N2X 3A1
Telephone: (1) (514) 8454419
Categories: D8

Ealing Scientific Ltd.
P.O. Box 238
Pointe Claire-Dorval
Quebec H9R 4N9
Canada
Telephone: (1) (514) 6311807
Categories: D2

DENMARK

New Dimensions Laser Systems A
Magstraede 5
1204 Copenhagen K
Denmark
Telephone: (45) (1) 326366
Categories: D2

EGYPT

Newport: Scientific & Trading
P.O. Box 114
Heliopolis West
Cairo
Egypt
Telephone: (20) (2) 831 025
Categories: D2 D3 D4

FRANCE

Ealing SA.R.L.
1030 Boulevard Jeanne D'Arc
59500Douai
France
Telephone: (33) (27) 884 865
Categories: D2

Soro Electro-Optics, S.A.
26 rue Berthollet
94110 Arcueil
France
Telephone: (33) (1)6571283
Categories: D4

Dept of Spectrophysics
Univ. Louis Pasteur
7 rue de L'Universite
67000 Strasbourg
France
Categories: D5

EQUIPMENT & SUPPLIES

D1=Film	**D5**=Labware	**D9**=Positioning Equipment
D2=Lasers	**D6**=Photochemistry	**D10**=Spatial Filters
D3=Mirrors	**D7**=Light Meters	**D11**=Pinholes
D4=Lenses	**D8**=Shutters	**D12**=Vibration Isolation Tables

ITALY

Coherent S.RL
Residenza Mestieri
Milano 2
20090 Segrate
Italy
Telephone: (39) (2) 2138 905
Categories: D2

Ealing Italia
Via Mario Greppi 5
28100 Novara
Italy
Telephone: (39) (0321) 28065
Categories: D2

GSC Laser Ltd.
Via Garibaldi 7
10122 Torino
Italy
Telephone: (39) (011) 555075
Categories: D2

JAPAN

Fuji Photo Optical Co. , Ltd.
Attn: Takayuki Saito
No. 324,1-Chrome
Vetake-Machi/ Omiya
Japan
Telephone: (81)(486) 630111
Categories: D3 D4

Newport: Kyokuto Boeki Kaisha
7th Floor, New Otemachi bldg.
2-1, 2-Chome, Otemachi
Chiyoda-ku, Tokyo 100-91
Japan
Telephone: (81)(3) 244 3511
Categories: D2 D3 D4

Hidetoshi Katsuma
Department of Electro Photo Op
Tokai University
1117 Kitakaname Hiratsuka City
Kanagawa 259-12,Japan
Categories: D9

NETHERLANDS

Coherent, B.V.
Meenthof15
1241 C.P. Kortenhoef
The Netherlands
Telephone: (31) (35) 62504
Categories: D2

Dutch Holographic Laboratory
Kanaaldyk Noord 61
Eindhoven 5642JA
The Netherlands
Telephone: (31)(40) 817250
Contact: Walter Spierings, Dir.
FAX: (31) (40) 814865
Categories: A1 A3 A4 A5 A6 A7 A8 A9 A10 B3 D1

Que Sera Sera
P.O. Box 29
9700 AA Groningen
The Netherlands
Telephone: (31)(050) 140417
Contact: H.T. Vogd
FAX: (31)(050) 144142
*Categories:*A12 B3 C4 D3 D4

SWEDEN

Institute of Physics, LTH
Box 118
22100 Lund
Sweden
Telephone: (46)(46) 107656
Categories: C5 D5

More than just a useful catalog...

**We stock
Agfa and Kodak
Holographic Film**

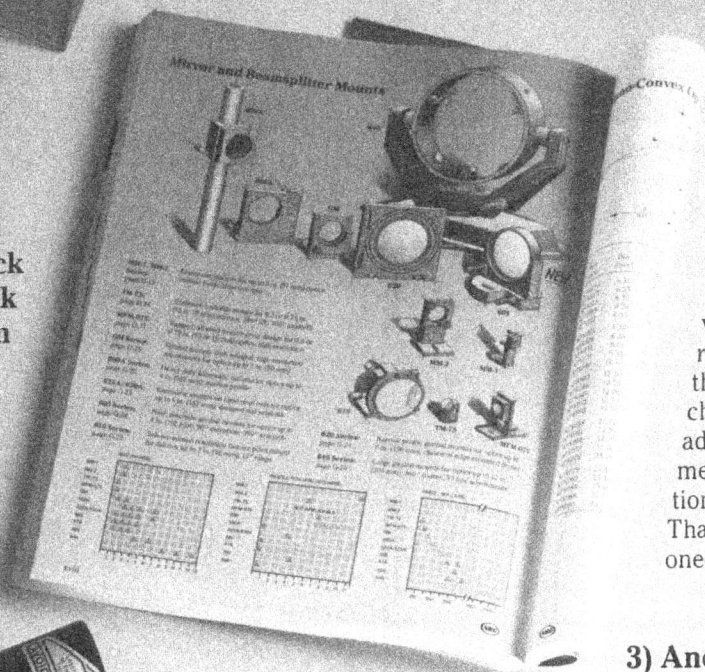

2) It offers just about everything you need.

When you reach for the Newport Catalog, you'll find what you need to make your research more productive: the best in optics, the widest choice in vibration isolation, advanced electro-optic instruments, and the broadest selection of precision positioners. That's the convenience of one-stop shopping.

1) It's informative and easy to use.

New *Selection Guides* let you zoom-in right to the best product for your application. Each product is described in detail, with helpful applications information, complete drawings and specifications, and cross-references to help you make an informed choice. Compatibility is designed-in across the board, so everything's easy to put together the way you want it. And our product line grows with your needs: for example, our new fiber optic and vibration isolation products help you keep pace.

3) And it's backed by the best service in the industry.

Call us! Our technical staff is always happy to assist you in your applications. And quick shipment, quality construction, and enduring compatibility all add extra value to the Newport Catalog. It's free—call or write for your copy.

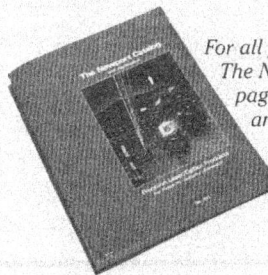

For all your laser/optic needs: The Newport Catalog. Over 400 pages of precision products and helpful technical information. It's free—call or write for your copy today.

714/965-5406

Newport Corporation
18235 Mt. Baldy Circle
Fountain Valley, CA 92708
Europe: Newport GmbH, Ph. 06151-26116
U.K.: Newport Ltd., Ph. 05827-69995

Newport

©1988 Newport Corporation

EQUIPMENT & SUPPLIES

D1=Film	**D5**=Labware	**D9**=Positioning Equipment
D2=Lasers	**D6**=Photochemistry	**D10**=Spatial Filters
D3=Mirrors	**D7**=Light Meters	**D11**=Pinholes
D4=Lenses	**D8**=Shutters	**D12**=Vibration Isolation Tables

UNITED KINGDOM

AG Electro-Optics Ltd.
29 Forest Road
Tarporley, Cheshire CW6 OHX
England
United Kingdom
Telephone: (44) (8293) 3305
Categories: D2 D4 D7 DIO

Agfa-Gevaert Ltd.
27 Great West Road
Brentford, Middlesex
TW8 9AX England
United Kingdom
Telephone: (44) (1) 560 2131
Categories: Dl

Coherent, Ltd.
Science Park
Milton Road
Cambridge, CB44BH
England
Telephone: (44) (223) 68501
Categories: D2

CSI
7 Meadowfield Park South
Stocksfield, Northumberland
NE43 7QA, England
United Kingdom
Telephone: (44)(661) 842741
Categories: D3

Datasights Ltd.
Alma Road, Ponders End
Enfield, Middlesex
EN3 7BB, England
United Kingdom
Telephone: (44)(1) 8054151
Categories: D3

Davin Optical Ltd.
Reliant House, Oakmere Mews
Potters Bar, Hertfordshire
EN69XX, England
United Kingdom
Telephone: (44)(707) 44445
Categories: D3

Ealing Electro-Optics
Greycaine Road
Watford, Hertfordshire
England WD2 4PW
United Kingdom
Telephone: (44) (923) 42261
Categories: D2

Electro Optics Developments Lt
Howards Chase
Pipps Hill Industrial Estate
Basildon, Essex
England, SS14 3BE, UX
Telephone: (44)(268) 20511
Categories: D3 D9 D10 D11 D12

Galvoptics Ltd.
HaIVeyRoad
Basildon, Essex
SS13 lES, England
United Kingdom
Telephone: (44)(0268) 728077
Contact: RD. Wale
FAX: (44) (0268) 590 445
Categories: D3 D4

Holofax Limited
Netherwood Road
Rotherwas Industrial Estate
Hereford HR2 6JZ
England, United Kingdom
Telephone: (44) (432) 278400
Categories: A8 D6 D12

EQUIPMENT & SUPPLIES

D1=Film	**D5**=Labware	**D9**=Positioning Equipment
D2=Lasers	**D6**=Photochemistry	**D10**=Spatial Filters
D3=Mirrors	**D7**=Light Meters	**D11**=Pinholes
D4=Lenses	**D8**=Shutters	**D12**=Vibration Isolation Tables

Howard Smith Precision Optics
61 Lancaster Road
New Barnett, Hertfordshire
EN4 BA5, England
United Kingdom
Telephone: (44)(1) 441 7878
Categories: D3 D4

Ilford Limited
Attn: Sabine Preston
30 Buckingham Gate
London, SWIE 6LB
England, United Kingdom
Telephone: (44) (1) 8285676
Categories: D1

Kendall Hyde Ltd.
Kingsland Industrial Park
Stroudley Road
Basingstoke, Hants.
RG24 OUG, England, UX
Telephone: (44) (0256) 840830
Contact:. M.D. Kendall
FAX:. (44) (0256) 840443
Categories: D3

Laser Instrumentation Ltd.
Unit 1 Bear Court
Daneshill East, Basingstoke
Hampshire RG24 OQT
England, United Kingdom
Telephone: (44) (256) 469 572
Categories: D2

Maplin Electronics Supplies Ltd
P.O. Box 3
Rayleigh, Essex
&S68LR
England
Telephone: (44) (702) 554 000
Categories: D3

METAX LTD
P.O. Box 315
Gladstone Road
Croydon, Surrey
England CR9 2BL
Telephone: (44)(1) 6896821
Categories: D2

Radio Shack/Tandy Corporation
Bilston Road,
Wednesbury
West Midlands, WSIO 7JN
England, United Kingdom
Categories: D7

Siemens Ltd.
Siemens House
Windmill Road
Sunbury-on-Thames
Middlesex, TWI6 7HS, England
Categories: D2

Spectra Physics Ltd.
17 Brick Knoll Park
St. Albans, Hertfordshire
ALI 5UF, England
United Kingdom
Categories: D2

Vinten Electro Optics Ltd.
Unit 28 Ashfield Way
Whetstone, Leicester
LE8 3NU, England
United Kingdom
Telephone: (44)(533) 867110
Categories: D3

EQUIPMENT & SUPPLIES

D1=Film	**D5**=Labware	**D9**=Positioning Equipment
D2=Lasers	**D6**=Photochemistry	**D10**=Spatial Filters
D3=Mirrors	**D7**=Light Meters	**D11**=Pinholes
D4=Lenses	**D8**=Shutters	**D12**=Vibration Isolation Tables

Wentworth Laboratories Ltd.
Sunderland Road
Sandy, Bedfordshire
SGI9 IRB, England
United Kingdom
Telephone: (44)(0767) 81221
Contact: M.F. Horgan
FAX: (44) (0767) 291951
Categories: Dl2

UNITED STATES OF AMERICA

Agfa-Gevaert Inc.
100 Challenger Road
Ridgefield Park, NJ
07660
Telephone: (1) (201) 4402500
Contact:. Mark Redzikowski
FAX: (1) (201) 3424742
Categories: DI

Apollo Lasers Inc
20415 Nordhoff Street
Chatsworth, CA 91311
Telephone: (1) (818) 4073016
Contact:. Ralph Page
FAX: (1)(818) 8825900
Categories: D2 .

City Chemical
132 West 22nd Street
New York, NY
10011
Telephone:. (1) (212) 929 2723
Categories: D6

Coherent Inc.
3210 Porter Drive
Palo Alto, CA
94304
Telephone In USA: (1) (800) 5273786
Telephone: (1) (415)493 2111
Categories: D2

Continental Optics
15 Power Drive
Hauppauge, NY
11788
USA.
Telephone: (1) (516) 5823388
Categories: Dl D3D4

Coulter Optical Company
P.O.BoxK
54121 Pinecrest Road
Idyllwild, CA
92349
Telephone: (1) (714) 6592991
Categories: D3

CVI Laser Corp.
200 Dorado Place SE
P.O. Box 11308
Albuquerque, NM
87192
Telephone: (1)(505) 296 9541
Categories: D2

Ealing Corporation
Pleasant Street
South Natick, MA
01760
USA.
Telephone:. (1) (617) 6557000
Categories: D3

Eastman Kodak
Charles Lysogorski
1447 St Paul Street
Rochester, NY
14650
Telephone: (1) (716) 2536795
Categories: D 1 D6

EQUIPMENT & SUPPLIES

D1=Film
D2=Lasers
D3=Mirrors
D4=Lenses

D5=Labware
D6=Photochemistry
D7=Light Meters
D8=Shutters

D9=Positioning Equipment
D10=Spatial Filters
D11=Pinholes
D12=Vibration Isolation Tables

Electro Optical Industries, Inc.
859 Ward Drive
Santa Barbara, Ca.
93111
USA
Telephone: (1) (805) 964 6701
Categories: D3

G.M. Vacuum Coating Lab, Inc.
882 Production Place
Newport Beach, CA
92663
USA
Telephone: (1)(714) 6425446
Categories: D3

Holographic Film Company
361 West Broadway
New York, NY
10013
USA
Telephone: (1)(212) 431 3420
Categories: D 1

HoloSpectra
7742-B Gloria Avenue
Van Nuys,CA
91406
Telephone: (1) (818) 9942577
Contact: W Arkin
FAX: (1)(818) 9944709
Categories: Al A4 A5 A6 A 7 A8 A9 AI2 B3 B8 D2
D3 D4 D5 D6 D7 D8 D9 DlODll DI2

Hughes Aircraft Co.
Laser Products
6155 EI Camino Real
Carlsbad, CA
92008
Telephone: (1) (714) 438 9191
Categories: D2

Illinois Valley Magnetic Resonance
Director of Resource Development
4005 Progress Boulevard
Peru,IL 61354
Telephone: (1) (815) 223 8674
Contact: Dr. John L. Mori
Categories: D2

Integraf
745 North Waukegan Road
Lake Forest, IL
00045
Telephone: (1) (312) 2343756
Contact: Anna Wong
Categories: B2 B3 B4 B7 C3 C6 c7 DI D3 D6

International Dikrotek, Inc.
755 South 200 West
Richmond, UT
84333
Telephone: (1) (801) 2582458
Categories: Dl

Ion Laser Technology
263 Jimmy Dolittle Road
Salt Lake City, UT
84116
USA
Telephone: (1) (801) 5371587
Categories: D2

Jodon Incorporated
62 Enterprise Drive
Ann Arbor, MI
48103
USA
Telephone: (1) (313) 761 4044
Contact: Preston Miller
FAX: (1) (313) 7613322
Categories: Dl D2 D3 D4 D5 D6 D7 D9 D10 D11 D12

EQUIPMENT & SUPPLIES

D1=Film	**D5**=Labware	**D9**=Positioning Equipment
D2=Lasers	**D6**=Photochemistry	**D10**=Spatial Filters
D3=Mirrors	**D7**=Light Meters	**D11**=Pinholes
D4=Lenses	**D8**=Shutters	**D12**=Vibration Isolation Tables

Kaiser Optical Systems, Inc.
P.O. Box 983
Ann Arbor, MI
48106
Telephone: (1) (313) 665 8083
Categories: D3 D4

Lambda/Ten Optics
One Lyberty Way
Westford, MA
01886
Telephone: (1)(508) 6928140
Contact: George Olmsted
FAX: (1)(508) 6929416
Categories: D3

Laser Institute Of America
Education Director
5151 Monroe Street-Suite 102W
Toledo,OH
43623
Telephone: (1)(419) 8828706
Categories: C12 D2

Laser Resale Inc.
54 Balcim Road
Sudbury,MA
01776
USA
Telephone: (1)(617) 4438484
Categories: D2

Metrologic Instruments, Inc.
P.O. Box 307
143 Harding Avenue
Bellmawr, NJ
08031
Telephone: (1) (609) 9330100
Categories: D3

Mitutoyo Measuring Instruments
18 Essex Road
Paramus, NJ
07652
USA
Telephone:(201)~525
Categories: D5 D7

Newport Research Corporation
18235 Mount Baldy Circle
Fountain Valley, CA
92708
USA
Telephone: (1)(714) 9639811
Contact: Technical Suppt
Categories: B4 Dl D4 D5 D9 D 12

Norland Products, Inc.
695 Joyce Kilmer Avenue
P.O. Box 145
North Brunswick, NJ
08902
Telephone: (1) (201) 545 7828
Contact: Jean Spalding
FAX: (1)(201) 5459542
Categories: D6

Optics Plus Inc.
1369 East Edinger Ave.
Santa Ana, CA
92705
USA
Telephone: (1) (714) 9721948
Categories: D2 D3 D4

Optical Purchasing Directory
Optical Publishing Co. Inc.
P.O. Box 1146
Berkshire Common
Pittsfield, MA 01201
Telephone: (1) (413) 499 0514
Categories: D2 D3 D4

EQUIPMENT & SUPPLIES

D1=Film	**D5**=Labware	**D9**=Positioning Equipment
D2=Lasers	**D6**=Photochemistry	**D10**=Spatial Filters
D3=Mirrors	**D7**=Light Meters	**D11**=Pinholes
D4=Lenses	**D8**=Shutters	**D12**=Vibration Isolation Tables

Electro Optical Specialist
Mr. Robert G. Stohl
SJ. Fed.Bldg-US Court House
280 South 1st Street, Rm 2062
SanJose, CA 95113
Telephone: (1) (408) 291 7893
Contact: Re:Laser Regulations
Categories:. D2

Optimation
289 Main Street-Suite 4
Salem,NH
03079
USA
Telephone: (1) (603) 434 2346
Categories: D12

Photographic Formulary
P.O. Box 3105
Missoula, MT
59806
Categories: D6

PMS Electro-Optics
1855 South 57th Court
Boulder,CO
80301
Categories: D2 D3 D4

Polaroid Corporation
Research Laboratories
750 Main Street#IA
Cambridge, MA 02139
Telephone:. (1) (617) 5773509
Contact;J.Cowan
Telephone: (1)(617) 577 4169
Telephone:.(l) (617) 5773805
Categories:. A12 D1 D6

Radio Shack-Tandy Corporation
Fort Worth, TX 76102
Telephone: (1)(817) 390 3011
Categories: D7

Science & Mechanics Instrument
605 East 59th Street
Brooklyn, NY
11234
Telephone: (1) (718) 5313381
*Categories:*D7 D8

Shipley Chemical Co.
1457 McArthur
Whitehall, PA
18052
Contact: Stu Price
Telephone: (1) (800) 345 3100
Categories: D6

Spectra Lumicon
c/o Simon & Associates
20 Sunnyside
Mill Valley, CA
94941
Telephone: (1)(415) 381 0835
Categories: D2 D3

Spectra-Physics
Laser Products Division
1250 West Middlefield Road
Mountain View, CA
94042
Telephone:(1) (415) 961 2550
Contact: Steve Anderson
FAX: (1)(415) 9694084
Categories: D2 D3 D4

Spectrogon AB
550 County Ave.
Secaucus, r.rr
07094
USA
Telephone: (1) (201) 867 4888
Categories: D3

EQUIPMENT & SUPPLIES

D1=Film	**D5**=Labware	**D9**=Positioning Equipment
D2=Lasers	**D6**=Photochemistry	**D10**=Spatial Filters
D3=Mirrors	**D7**=Light Meters	**D11**=Pinholes
D4=Lenses	**D8**=Shutters	**D12**=Vibration Isolation Tables

WEST GERMANY

Ealing GmbH
Bahnhofstr. 8
Postfuch 1226
Hochst
Federal Republic of Germany
Telephone: (49) (6131) 3909
Categories:. D2

Newport:Carl Baasel Lasertechnic
Sandstrasse 21
8000 Munchen 2
Federal Republic of Germany
Telephone: (49) (89) 521100
Categories: D2 D3 D4

Spectra-Physics GmbH
Alsfelder Strasse 12
0-6100 Darmstadt
Federal Republic ofGerrnany
Telephone: (49) (6151) 7081
Categories: D2

Holography News™

The International Business Newsletter of the Holography Industry

CHAPTER
8

BIBLIOGRAPHY

This bibliography represents a listing of holography sources from one of the largest university library systems in the USA, The University of California. Over 200 titles on Holography are catalogued here, assembled from the nine campuses throughout the state.

The first section includes books, seminar publications, and dissertations in English. The second section consists of periodicals in English. The last section is a listing of books, seminar publications, and periodicals in foreign languages as well as a brief listing of miscellaneous items of interest not found in the UC libraries.

Using the Interlibrary Loan system at your local public or private library in the USA, you can borrow any book from this listing using the call number cited. The University may place some titles on reserve.

The abbreviations beside the call numbers stand for the specific University of California campus where the title is found:

LBL-Lawrence Berkeley Labs, Berkeley
UCB-Berkeley
UCD-Davis
UCI-Irvine
UCLA-Los Angeles
UCR-Riverside
UCSB-Santa Barbara
UCSC-Santa Cruz
UCSD-San Diego
CSL-California State Library, Sacramento
SRLF-Southern Regional Library Facility

Abramson, Nils H. *The making and evaluation of holograms.* London: Academic Press, 1981.
    UCB Engin TA1540.A37
    UCD Phys Sci TA1540.A37
    UCI Main Lib TA1540.A37
    UCSB Library TA1540 .A37 Sci-Engrg
    UCSC Science TA1540.M7
    UCSD S&E TA1540.A37

Acoustic imaging: cameras, microscopes, phased arrays, and holographic systems. Glen Wade, ed. New York: Plenum Press, 1976.
    UCD Phys Sci TA1550.A26
    UCI Main Lib TA1550 .A26
    UCR Phy Sci TA1550 A261976
    UCSB Library TA1550 .A26 Sci-Engrg
    UCSC Science TA1550A26

Aldridge, Edward E. *Acoustical holography.* Watford: Merrow Publishing Co. Ltd., 1971. Series title: *Merrow monographs, practical science series 1.*
    UCI PhySciLib QC244.5 .A43
    UCLA Engr/Math QC 244.5 A365a
    UCSD S&E QC244.5 A43

Anderson, John. *Holography.* Tempe: Art Dept., Arizona State University, 1979. Series title: Northlight; no. 11.
    UCR Rivera TR7 .N6 no.ll

Applications of holography: January 21-23, 1985, Los Angeles, California. Lloyd Huff, chair, ed. Bellingham, WA: SPIE-the International Society for Optical Engineering, 1985.

   Series title: Proceedings of SPIE-the International Society for optical Engineering; v. 523.
      UCB Engin TA1540 .A66 1985
      UCD PhysSci TS510 .S63v.523
      UCI Main Lib TS510 .P632 v. 523
      UCLA Engr/Math TA 1540A661985
      UCSB Library TAl540 A671985 S&E
      UCSD S & E TA1540 A671985

Applications of lasers to photography and information handling; proceedings, two-day seminar. Richard D. Murray, ed.Washington, DC: Society of Photographic Scientists and Engineers, 1968.
      UCLA Engr/Math TK 7872 L3M96
      CSL Main Lib TK 7871 .3 A8 General Coll

Beiser, Leo. *Holographic scanning.* New York: W ley, 1988. Series title: *Wiley series in pure and applied optics.*
      UCD Phys Sci TK7882.S3 B45 1988
      UCI Main Lib TK7882.s3 B45 1988
      UCLA Engr/Math TK 7882 S3 B45 1988
      UCSD S & E TK7882.S3 B45 1988

Berley, Lawrence F. *Holographic mind, holographic vision: a new theory of vision in art and physics.* 1st ed. Bensalem, PA: Lakstun Press, 1980.
      UCB Optom QC449 .B471980
      UCSC Science QC449.B4 7 1980

Berner, Jeff. *The holography book.* New York: Avon Books,1980.
      UCI Main Lib N6494.L3 B47
      UCSD Central N6494.L3 B47

Brcic, Vlatko. *Application of holography and hologram interferometry to photoelasticity: lectures held at the Department for Mechanics of Deformable Bodies.* 2d ed. Wien: Springer-Verlag, 1974. Series title: *Courses and lectures ; no. 7.*
      UCLA Engr/Math QC 449 B7391972

Butters, John N. *Holography and its technology.* London: P. Peregrinus,1971; Published on behalf of the Institution of Electrical Engineers. Series title: *Institution of Electrical Engineers* l. E. E. monograph, series, 8.
      UCD Phys Sci TK5 .142 no.8
      UCI Main Lib TAI540 .B87
      UCLA Engr/Math QC 449 B982h
      UCSB Library QC449 .B88 Sci-Engrg

Cadig Liaison Centre. *Reference Library. A compendium of Cadig bibliographies: Metrication, Fluidics, Explosive techniques in engineering, Holography, Carbon fibres.* Coventry (Warwickshire) : Cadig Liaison Centre, 1970.
      UCD Phys Sci Z7144.W4 C3
      UCSB Library Z7144.W4 C3 Sci-Engrg Ref

Cathey, W. Thomas. *optical information processing and holography.* New York:Wiley,1974. Series title: *Wiky series in pure and applied optics.*
      UCB Physics TA1630.C371
      UCD Phys Sci TA1630.C37
      UCI Main Lib TA1630 .C37
      UCR Phy Sci TA1630 .C37 1974
      UCSB Library TA1630 .C37 Sci-Engrg
      UCSD S & E TAI630 .C37

Caulfield, H. J. and Sun Lu. *The applications of holography.* New York, Wiley-Interscience, 1970. Series title: *Wiley series in pure and applied optics.*
      LBL Bldg 50 QC449 .C372 1970
      UCB Moffitt QC449 .C37
      UCD Phys Sci QC449 .C37
      UCI PhySciLib QC449 .C37
      UCR Phy Sci QC449 .C37 1970
      UCSB Library QC449 .c37 Sci-Engrg
      UCSC Science QC449.C37
      UCSD S & E QC449 .C37
      UCSD Scripps QC449 .C37
      CSL Main Lib QC 449 c37 General Coll

Ceccon, Harry L. *Holographic techniques for nondestructive testing of tires.* Washington, D.C: National Highway Traffic Safety Administration, 1972.
      UCB Trans N.S. 72-72

Centerbeam. Otto Piene and Elizabeth Goldring, eds. Introduction by Lawrence Alloway. Cambridge, MA: Center for Advanced Visual Studies, Massachusetts Institute of Technology, 1980.
> UCD Main Lib NB1272 .C45
> UCI Main lib NB1272.C46
> UCR Rivera NB1272.C46

Chuguy, Yu. V. and N. T. Kolesova. *Bibliography on holography*, 1971-1972. Translated from Russian. London: Scientific Information Consultants Ltd, 1976.
> UCLA Engr/Math ZQC 449 C471E 1971/7

Coello-Vera, Agustin Elias. *"Scanned acoustic imaging in the ocean : a study of holographic-like systems and their limitations"*. 1978.
> UCSB Library TA1550 .C63 Sci-Engrg

Coherent optical processing: seminar, August 21-22, 1974, San Diego, CA. Palos Verdes Estates, CA: Society of Photo-optical Instrumentation Engineers, 1975. Series title: *Society of Photo-optical Instrumentation Engineers Seminar proceedings* ; v. 52.
> UCI Main lib TS510 .P632v. 52
> UCLA Engr/Math TA 1630 C66 1974
> UCSB Library TA1630 .C63 Sci-Engrg
> UCSD S&E TA1630.C63

Coherent optics in mapping: tutorial seminar and technology utilization program, March 27-29, 1974, Rochester, N.Y. N. Balasubramanian, Robert D. Leighty, eds. Jointly sponsored by American Society of... Palos Verdes Estates, CA: SPIE,1974.
> Series title: *Society of Photo-optical Instrumentation Engineers Proceedings;* v. 45.
> UCD Phys Sci TR810 .B3
> UeI Main lib TS5lO .P632 v. 45
> UCSB Library TA1542 .C63 Sci-Engrg
> UCSD S & E TA1542 .C63

Collier, RobertJacob, C. B. Burckhardt and L.H. lin. *Optical holography*. New York: Academic Press, 1971.
> LBL Bldg 50 QC449 .C699 1971
> LBL Bldg 90 QC449 .C699 1971 LongLoan
> LBL Donner QC449 .C699 1971
> UCB Physics QC449 .C64
> UCD Phys Sci QC449 .C64
> UCI PhySciIib QC449 .C64
> UCLA Biomed QC 449 C6910 1971
> UCLA Engr/Math QC 449 C690
> UCR Phy Sci QC449.C641971
> UCR Rivera QC449.C64
> UCSB library QC449 .C64 Sci-Engrg
> UCSC Science QC449.C64
> UCSD S & E QC449 .C64
> UCSD Scripps QC449 'C64
> UCSF General QC449.C641971

Conference on Fourier Optics, Lasers and Holography, Mysore, 1971. Proceedings of the Conference on Fourier Optics, Lasers and Holography, Mysore, November 11-15, 1971. Madras, India: Institute of Mathematical Sciences, 1971. Series title: *Matscience report;* 77.
> UCD Main lib QA1 .M92 no.77
> UCI PhyScilib QA3 .M3 no.77
> UCSB library QA4A1 M3 no.77 Sci-En
> CRL GenCollec 1).10493 Type EXPLAIN
> CRL for borrowing information.

Conference on Holographic Instrumentation Applications, 1970: Ames Research Center. Holographic instrumentation applications. Prepared by NASA Ames Research Center. Boris Ragent and Richard M. Brown, eds. Washington, DC: Scientific and Technical Information Division, National Aeronautics and Space Administration,1970. Series title: *NASA SP; 248.*
> UCLA Engr/Math QC 449 C76h 1970

Conference on Holography and Optical Filtering, 1972: Marshall Space Flight Center. Holography and optical filtering; proceedings. Washington, Scientific and Technical Information Office, National Aeronautics and Space Administration,1973. Series title: *United States. National Aeronautics and Space Administration* NASA SP -299.
> UCI Main lib TA1540 .C66 1971
> UCLA Engr /Math QC 449 C77h 1971
> UCSB Library TAl540 .C66 1971 Sci-Engrg
> UCSC Science TA1540.C661971

Dallas, William John. *"Computer holograms: improving the breed"*. 1971.
UCSD Central QC 3.6 .D244 Spec ColI Archives Diss
UCSD S & E QC 3.6 .D244

Defense Documentation Center (U.S.) *Holography: a DDC bibliography*, January 1970-September 1972. Alexandria, VA: Defense Documentation Center, Defense Supply Agency, 1973.
UCSB Library QC449 .U45 1973 Sci-Eng Ref

Denisiuk, IU. N. *Fundamentals of holography*. Translated from the Russian by Alexander Chubarov. Rev. from the 1978 Russian ed. Moscow: Mir,1984.
UCD Phys Sci TA1540 .D4613 1984

DeVelis, John B. and George O. Reynolds. *Theory and applications of holography*. Reading, MA: AddisonWesley Pub. Co., 1967.
UCD Phys Sci QC449 .D4
UCI PhySciLib QC449 .D4
UCLA Engr /Math QC 449 D492t
UCR Phy Sci QC449 .D4
UCSB Library QC449 .D4 Sci-Engrg
UCSC Science QC449.D4
UCSD S & E QC449 .D4
CSL Main Lib QC 449 D4 General Call

Developments in holography: seminar-in-depth; proceedings. Brian J. Thompson and John B. DeVelis,eds. Redondo Beach, CA: Society of Photo-optical Instrumentation Engineers, 1971. Series title: *Society of Photo-optical Instrumentation Engineers S.P.I.E.seminar proceedings*, v. 25.
LBL Donner QC449 .D489 1971
UCD Phys Sci QC449.D421971
UCLA Engr/Math QC 449 S471d 1971
UCSB Library QC449 .D43 1973 Sci-En
UCSD S & E QC449 .D43

Dowbenko, George. *Homegrown holography*. Garden City, N.Y.: Amphoto, 1978.
UCI PhySciLib QC449 .D68
UCSB Library QC449 .D68 Sci-Engrg
UCSD Central QC449 .D68 Oversize

Dudley, David D. *Holography; a survey*. Washington, DC: Technology Utilization Office, National Aeronautics and Space Administration,1973. Series title: *NASA SP-5118*.
UCB Engin TAI540 .D84
UCB Optom TAI540 .D84
UCI Main Lib TA1540 .D8
UCLA Engr/Math QC 449 D865h
UCSB Library TA1540 .D8 Sci-Engrg

Dzekov, Tomislav Angel. *"Microwave holographic imaging of aircraft with space borne illuminating source"*. 1976.
UCSB Library TA1552 .D94 1976a SciEngrg

Micro-film Eichert, Edwin S. and Alan H. Frey; Randomline, Inc. *Holography in driver education, training, testing, and research*. Washington, DC : National Highway Traffic Safety Administration, 1978. Series title: *United States. National Highway Traffic Safety Administration Report*; no. DOT HS-803 035.
UCB Trans N.S.77-384

Eichler, H. j., P. Gunter, and D.W. Pohl. *Laserinduced dynamic gratings*. Berlin: SpringerVerlag, 1986. Series title: *Springer series in optical sciences*; v. 50.
UCB Physics QC449 .E34 1986
UCD Phys Sci QC450.s65 v.50
UCI PhySciLib QC417 .E35 1986
UCLA Engr/Math QC 417 E32 1986
UCR Phy Sci QC417.E35 1986
UCSB Library QC417 .E35 1986 Sci-En
UCSC Science QC417.E381986
UCSD S&E QC417.E4251986

Electro-optics/laser international '80 UK. Brighton, 25- 27 March 1980 :conference proceedings. H.G. Jerrard, ed. Conference organized by Kiver Communications Ltd. Guildford, Surrey, England: IPC Science and Technology Press, 1980.
UCB Engin TA1505 .E4
UCD Phys Sci TA1505 .E4
UCI Main Lib TA1505 .E4
UCLA Engr/Math TA 1505 E381980
UCSC Science TA1505 .E431980

Elliott, John Douglas. *"Computer simulation of the holographic image degradation due to transmission of the signal through a random noise media"*. 1971.

SRLF D 0000348540 Type EXPlAIN SRLF for borrowing information.

Engineering Applications of Holography Symposium, 1972: Los Angeles:Proceedings. Redondo Beach, CA: Society of Photo-optical Instrumentation Engineers, 1973.

UCD Phys Sci TA1501 .E53 1972
UCI Main Lib TA1501 .E53 1972 Oversize
UClA Engr/Math TA 1540 E58p 1972
UCSB Library TA1501 .E53 1972 Sci-Engrg
UCSD S&E TAI50l.E531972

The Engineering uses of coherent optics: proceedings and edited discussion of a conference held at the University of Strathclyde, Glasgow, 8-11 April,1975 . Organised by the University, in association with the... Cambridge, Eng. : Cambridge University Press, 1976.

UCI Main Lib TAI505.E531976
UCLA Engr/Math TA 1520 E57 1975
UCSB Library TAI505.E531976Sci-Engrg
UCSD S&E TAI505.E531976

The Engineering uses of holography. Elliot R. Robertson and James M. Harvey, eds. Cambridge, Eng.: Cambridge University Press, 1970.

UCD Phys Sci TA165 .E54
UCI Main Lib TA165 .E54
UCLA Engr /Math QC 449 E5 75 1968
UCR Phy Sci TAI65 .E541968
UCSB Library TA165 .E54 Sci-Engrg
UCSD S & E TAI65 .E54
CSL Main Lib T A 165 E54 General ColI

Eu, James Kim-Tzong. "Studies in spatial filtering". 1974.

UCSD Central QC 3.6 .E9 Spec Coli Archives Diss
UCSD S & E QC 3.6 .E9

*European Hybrid Spectrometer Workshop on Holography and High-Resolution Techniques,*1981: Strasbourg, France. *Photonics applied to nuclear physics, 1 / European Hybrid Spectrometer Workshop on Holography and High-Resolution Techniques, Strasbourg, Council of Europe, 9-12 November 1981.* Geneva: European Organization
for Nuclear Research, 1982. Series title:

CERN (Series) ; 82-01.
LBL Bldg 50 TAI522 .E91981
UCLA Physics QC 449 E971981

Evans, Evan Allen. "Quantitative method for accurate determination of red blood cell geometry".1970.

UCSD Central T3.6 .E917 Spec Coli Archives Diss
UCSD S&E T3.6.E917

Finch College, New York. Museum of Art. Contemporary Study Wing. "N dimensional space". Prepared by Ted McBurnett. Introd. by Elayne H. Varian. New York: 1970.

UCSC McHenry N6494.L3F5

Firth, Ian Mason. Holography and computer generated holograms. London, Mills and Boon, 1972. Series title: M & B monograph EE/ll.

UCI PhySciLib QC449 .F571972

Fischer, Wolfgang Klaus. Methods for acoustic holography and acoustic measurements. Newark, N,J. : [s.n.]' 1972.

UCSB Library QC244.5 .F57 Sci-Engrg

Flow visualization and aero-optics in simulated environments: 21-22 May 1987, Orlando, Florida. H. Thomas Bentley III, chair/ed. Sponsored by SPIE-the International Society for Optical Engineering. Bellingham, WA: SPIE-the International Society for Optical Engineering,1987. Series title: *Proceedings of SPIE-the International Society for Optical Engineering; v.788.*

UCB Engin TL567.W5 F581 1987
U CD Phys Sci TSSlO .S63 v.788
UCLA Engr/Math TA 357 F5631987
UCSB Library TA357 .F5311 1987 SciEn
UCSC Science QC370.5.S6 v.788
UCSD S & E TL567.W5 F58 1987

Francon, M. Hologrpahy. Expanded and revised from the French edition. Translated by Grace Marmor Spruch. New York: Academic Press, 1974.
  UCI PhyScilib QC449 .F713
  UCR Phy Sci QC449 .F713 1974
  UCSD S & E QC449.F713
  UCSD Scripps QC449.F713
  CSL Main Lib QC449 F713 General Coll

George, DaweelJoseph. *"Holography as applied to jet breakup and an analytical method for reducing holographic droplet data"*. 1972. SRLF D 0000427682 Type EXPlAIN SRLF for borrowing information. Graphics in motion : from the special effects film to holographics. John Halas, ed. New York: Van Nostrand Reinhold, 1984.
  VCB EnvDesign TR858 .G711984
  VCLA Theater TR 858 G7 1984

Handbook of optical holography. H. J. Caulfield,ed. Contributors, Gilbert April ... ret al.]. New York: Academic Press, 1979.
  UCB Engin TA1540.H36
  UCB Physics T A1540 .H36 Reserve
  UCD Phys Sci TAI540.H36
  UCI Main Lib TA1540.H36
  UCLA Physics TA 1540 H361979

Reference Area Shelved in Handbook Section.
  UCR Phy Sci TA1540 H36x, 1979
  UCSB Library TA1540 .H36 Sci-Engrg
  UCSC Science TAI540.H36
  UCSD S & E TA1540.H36

Hariharan, P. *Optical holography: principle's, techniques, and applications.* Cambridge: Cambridge University Press,1983. Series title: Cambridge monographs on physics.
  UCB Physics TA1540 .H37 1984
  UCD Phys Sci TAI540.H371984
  UCI Main Lib TA1540.H371984
  UCLA Physics TA 1540 H371984
  UCR Phy Sci TA1540.H371984
  UCSB Library TA1540 .H371984 Sci-En
  UCSC Science TA1540 .H37 1983
  UCSD S & E TAI540 .H371984

Hildebrand, B. P. and B. B.Brenden. An introduction to acoustical holography. New York: Plenum Press, 1972.
  UCD Phys Sci QC244.5 .H55
  UCI PhyScilib QC244.5 .H55
  UCR Phy Sci QC244.5 .H55 1972
  UCSB Library QC244.5 .H55 Sci-Engrg
  UCSD S & E QC2445 .H55

Holographic data nondestructive testing: October 4-8, 1982, Croatia Hotel de Luxe, Dubrovnik, Yugoslavia. Dalibor Vukicevic, chair,ed. Sponsored by the International Commission for Optics (ICO) [and] ... Bellingham, WA: SPIE-the International Society for Optical Engineering, 1 983. Series title: Proceedings of SPIE-the International Society for Optical Engineering; v. 370.
  UCB Engin TA417.2 .H64 1983
  UCD Phys Sci TS510 .S63 v.370
  UCI Main Lib TS5lO .P632 v. 370
  UCLA Engr/Math TA 1540 H65 1982
  UCR PhySci TAI540 .H651983
  UCSB Library TA1540 .H65 1983 Sci-En
  UCSD S & E TAI540 .H65 1983

"Holographic detection of intraocular pathology in the presence of cataracts: final report". By George O. Reynolds ... ret al.]. Burlington, MA: Technical Operations, 1974.
  UCB Optom RE451 .H6

Holographic nondestructive testing. Robert K. Erf, ed. New York: Academic Press, 1974.
  UCD Phys Sci TA417.2 .E73
  UCI Main Lib TA417.2.E73
  UCLA Engr/Math TA417.2 E671974
  UCR Phy Sci TA417.2 E73x, 1974
  UCSB Library TA417.2 .E73 Sci-Engrg
  UCSD S & E TA417.2.E73

Holographic nondestructive testing : status and comparison with conventional Methods : 23-24 January 1986, Los Angeles, California. Charles M. Vest, chair, ed. Presented in co-operation with American Association... Bellingham, WA: SPIE-the International Society for Optical Engineering, 1986. Series title: Proceedings of SPIE-the International Society for Optical Engineering; v. 604. Series title: SPIE critical reviews of technology series;15th.

    UCB Engin TA417.2 .H851 1986
    UCD Phys Sci TS5lO .S63 v.604
    UCI Main Lib TS510 .P632 v. 604
    UClA Engr/Math TA 417.2 H64 1986
    UCSB Library TA417.2 .H641986 Sci-En
    UCSC Science QC370.5.S6 v.604
    UCSD S & E TA1540 .H65 1986

The Holographic paradigm and other paradoxes: exploring the leading edge of science. Ken Wilber, ed. 1st ed. Boulder: Shambhala, 1982.

    UCB Main Q175 .H773 1982
    UCB Moffitt Q175 .H773 1982
    UCD Main Lib Q175 .H773 1982
    UCI Main Lib Q175 .H773 1982
    UClA College Q 175 H773 1982
    UCSB Library Q175 .H773 1982 Sci-En
    UCSC McHenry Q175.H773 1982
    UCSD Undergrad Q175 .H773 1982

Holographic recording materials. H. M. Smith,ed. Contributions by R. A. Bartolini... ret al.]. Berlin: Springer-Verlag, 1977. Series title: Topics in applied physics; v. 20.

    UCB Physics TA1542.H64
    UCD Phys Sci QC1.T66v.20
    UCI Main Lib TA1542.H64
    UCR Phy Sci TA1542.H641977
    UCSB Library TA1542 .H64 Sci-Engrg
    UCSC Science TA1542.H64
    UCSD S & E TAl542.H64

Holography. Redondo Beach, CA: Society of Photooptical Instrumentation Engineers, 1968. Series title: Society of Photo-optical Instrumentation Engineers S.P.I.E. seminar proceedings, v. 15. LBL Bldg 50 QC449. H7541968

    UCD Phys Sci QC449.H61968
    UCLA Engr/Math QC 449 H741
    UCSD S & E QC449 .H6

Holography: January 24-25, 1985, Los Angeles, CA. Lloyd Huff, chair,ed. Bellingham, WA: SPIE-the International Society for Optical Engineering, 1 985. Series title: Proceedings of SPIE-the International Society for optical Engineering; v. 532. Series title: SPIE critical reviews of technology series; 12th.

    UCB Engin TA1540 .H65 1985
    UCD Phys Sci TS510 .S63 v.532
    UCI Main Lib TS510 .P632 v. 532
    UClA Engr/Math TA 1540 H661985
    UCSB Library TA1540 .H661985 Sci-En
    UCSD S & E TA1540 .H661985

Holography; seminar-in-depth, May, 1968, San Francisco CA. B.G.Ponseggi and Brian J. Thompson,eds. Redondo Beach, CA: Society of Photo-optical Instrmentation Engineers, 1972. Series title: Society of Photo-optical Instrumentation Engineers Proceedings, v.15.

    UCSB Library QC449 .H6 1972 Sci-Engrg

IAroslavskii, L. P. and N. S. Merzlyakov. *Methods of digital holography.* Translated from Russian by Dave Parsons. New York: Consultants Bureau, 1980. LBL Bldg 90 TA1542 .127131977 Long Loan

    UCB Physics TA1542.12713
    UCD Phys Sci TA1542.12713
    UCI Main Lib TA1542 .12713
    UClA Physics TA 1542 127131980
    UCSB Library TA1542 .12713 Sci-Engrg
    UCSC Science TA1542.12713
    UCSD S & E TA1542.12713

Industrial and commercial applications of holography: August 24-25, 1982, San Diego, California. Milton Chang, chair,editor. Bellingham, WA: SPIE-the International Society for Optical Engineering,1983. Series title: Proceedings of SPIE-the International Society for optical Engineering; v. 353.

    UCB Engin TAl540.156
    UEN Phys &i TS510 .S63 v.353
    UeI Main Lib TS510 .P632v. 353
    UClA Engr/Math TA 1540 15271982
    UCSB Library TA1540 .15271983 &i-En
    UCSD S & E TA1540 .1527 1983

Industrial applications of holographic nondestructive testing: May 3-5, 1982, Brussels. J. Ebbeni, chair, ed. Sponsored by SPIE-the International Society for Optical Engineering ; with the support. .. Bellingham, WA: SPIE-the International Society for Optical Engineering, 1982. Series title: Proceedings of SPIE-the International Society for Optical Engineering; v. 349.

    UCB Engin TA417.2 .1541982
    UEN Phys &i TS510 .S63 v.349
    UCI Main Lib TS5lO .P632v. 349
    UClA Engr/Math TA 1540 1531982
    UCSB Library TA417.2 .1458 1982 &i-En
    UCSD S&E TA1540 .1531982

Industrial applications of laser technology: April 19-22, 1983, Geneva Switzerland. William F. Fagan, chair, ed. Bellingham, WA: SPIE-the International Society for Optical Engineering, 1983. Series title: *Proceedings of SPIE-the International Society for Optical Engineering;* v. 398.

    U CB Engin QC350 .S6 1983
    UCD Phys &i TA1555.1531983
    UCI Main Lib TS5lO .P632 v. 398
    UClA Engr/Math TA 1555 1531983
    UCSB Library TA1673 .15525 1983 &i-En
    UCSD S & E TA1555 .153 1983

International Commission for Optics. Congress, 10th: 1975: Prague, Czechoslovakia. Recent advances in optical physics : proceedings of the Tenth Congress of the International Commission for optics, August 25-29, 1975, Prague, Czechoslovakia. Bedrich Havelka and Jan Blabla,eds. Olomouc: Palacky University ; Prague: Society of Czechoslovak Mathematicians and Physicists, 1976.

    UCB Physics QC395.2.15619

International Conference on Applications of Holography and Optical Data Processing, 1976: Jerusalem, Israel. Applications of holography and optical data processing: proceedings of the international conference, Jerusalem, August 23-26, 1976. E. Marom, A. A. Friesem, and E. Wiener-Avnear, eds. 1st ed. Oxford: Pergamon Press, 1977.

    UCB Engin TA1542.15811976
    UCB Physics TA1542.15811976
    UEN Phys &i TA1542.1581 1976
    UCI Main Lib TA1542 .1581976
    UCR Phy&i TA1542.1581976
    UCSB Library TA1542 .1581976 Sci-En
    UCSD S&E TA1542.1581976

International Conference on Computer-generated Holography, 1983 : San Diego, CA. International Conference on Computer-generated Holography, August 25-26,1983, San Diego, CA: proceedings. Sing H. Lee, chair, ed. Bellingham, WA: SPIE-The International Society for Optical Engineering, 1 983. Series title: *Proceedings of SPIE-the International Society for Optical Engineering;* v. 437.

    UCB Engin TA1542 .16 1983
    UCD Main Lib TS510.S63v.437
    UeI Main Lib TS510 .P632v. 437
    UClA Engr/Math TA 15421591983
    UCSB Library TA1542 .1591983 Sci-En
    UCSD S & E TA1542.1591983

International Conference on Holography Applications, 1986: Peking, China. International Conference on Holography Applications: 2-4 July, 1986, Beijing, China. Dahang Wang, chair. Jingtang Ke, Ryszard J. Pryputniewicz, eds. Sponsored by COS-Chinese Optical Society ... Bellingham, WA: SPIE,1987. Series title: *Proceedings of SPIE-the International Society for Optical Engineering;* v. 673.

    UCB Engin TA1555 .1581 1987
    UEN Phys &i TS510 .S63 v.673
    UClA Engr/Math TA 15401581986
    UCSC &ience QC370.5.S6 v.673
    UCSD S & E TA1555.15841986

International Congress on High-Speed Photography, ll th:1974: Imperial College, London. High speed photography : proceedings oJ the eleventh International Congress on High Speed Photography, Imperial College, University oJ London, September 1974. P. J. Rolls, ed. London: Chapman & Hall: distributed in the USA by the Society of Photo-Optical Instrumentation Engineers, 1975.
> UCD Phys Sci TR593.1571974
> UCLA Engr /Math TR 593 161 1974
> UCSD S & E TR593 .163 1974

*International Congress on High Speed Photography, 12th:*1976: Toronto, Canada. Proceedings oJ the 12th International Congress on High Speed Photography (Photonics), Toronto, Canada, 1-7 August 1976. Martin C. Richardson. Bellingham, WA: Society of PhotoOptical Instrumen tation Engineers, 1977. Series title: SPIE v.97.
> UCB Engin TS510.s6v.97
> UCD Phys Sci TS5lO .s63v.97
> UCLA Engr / Math TR 593 161 1976
> UCSD S & E TR593 .163 1976

International Congress on High Speed Photography and Photonics, 13th: 1978: Tokyo. Proceedings of the 13th International Congress on High Speed Photography and Photonics-Tokyo, 20-25 August 1978. Shin-ichi Hyodo, ed. Tokyo: Japan Society of Precision Engineering; [New York] : distributed (outside Japan) by Society of Photo-Optical Instrumentation Engineers, 1979. Series title: SPIE v. 189.
> UCB Engin TS510.s6v.189
> UEN Phys Sci TS510 .s63 v.189
> UCSD S & E TR593.1631978

International Optical Computing Conference, 1974: Zurich. Digest oJ papers. New York, Institute of Electrical and Electronics Engineers, 1974.
> UCD Phys Sci TA1630 .1571974
> UCSD S&E TA1632.1571974

International Optical Computing Conference, 1975: Washington, D.C. Digest oJ papers: International Optical Computing ConJerence, April 23-25, 1975, Washington, D. C. Sponsored by the Computer Society of the Institute of Electrical and Electronic Engineers, in cooperation... New York: Institute of Electrical and Electronics Engineers, 1975.
> UCD Phys Sci TA1630 .157 1975
> UCLA Engr/Math TA 1630 I612d 1975
> UCSB Library TA1630 .157 1975 Sci-En
> UCSD S & E TA1632 .157 1975

International Symposium on Acoustical Holography. Acoustical holography. New York: Plenum Press, 1967.
> UCSC Science QC244.5.151967

International Symposium on Acoustical Holography. lst: 1967: Huntington Beach, CA. Acoustical holography; proceedings. New York: Plenum Press, 1969.
> UCI Biomed QC244.5 .A185 v.I-7, 1969

International Symposium on Acoustical Holography and Imaging, 7th: 1976: Chicago, IL. Recent advances in ultrasonic visualization. Lawrence W. Kessler, ed. New York: Plenum Press, 1977.Series title: *International Symposium on Acoustical Holography and Imaging Acoustical holography;* v. 7.
> UCLA Engr/Math QC 244.5 Al 161 1976

International Symposium on Acoustical Holography and Imaging, 8th, Key Biscayne, FL, 1978. Ultrasonic visualization and characterization. A. F. Metherell, ed. New York: Plenum Press,1980. Series title: *International Symposium on Acoustical Holography and Imaging Acoustical imaging;* v. 8.
> UCLA Engr/Math QC 244.5 Al 1611978

*International Symposium on Holography in Biomedical Sciences,*1973: New York. *Holography in medicine: proceedings of the International Symposium on Holography in Biomedical Sciences,* New York, 1973. Paul Greguss, ed. Guildford, Eng: IPC Science and Technology Press, 1975.
    UClA Biomed QC 449 161h 1973
    UCSD Biomed QC 449 I6161973h
    UCSD Central R857.H64 1571973
    UCSF General R857.H64 Al 1571975

International Workshop on Holography in Medicine and Biology, 1979: Munster, Germany. Holography in medicine and biology : proceedings of the International Workshop, Munster, Fed. Rep. of Germany, March 14-15, 1979. G. von Bally, ed. Berlin: SpringerVerlag, 1979. Series title:*Springer series in optical sciences;* v 18
    UCB Physics R857.H64 1581979
    UCD Phys Sci QC450 .S65 v.18
    UCI Biomed QT 34 161h 1979
    UClA Biomed TA 1540 H7541979
    UCLA Engr/Math R 857 H64 1571979
    UCR Bio-Ag R857.H64 158 1979
    UCSB Library R857.H64 1581979 Sci-En
    UCSC Science R857.H641581979
    UCSD S&E R857.H64H651979
    UCSF General R857.H64 1581979

Kallard, Thomas. *Holography; state of the art review,* 1969. New York: Optosonic Press,1969. Series title: State of the art review, 1.
    UCD Phys Sci Z7144.H6 K3
    CSL Main Lib Z7144 H6 K3 General ColI

____ . *Holography; state of the art review ...* 1970. Holography in1970: an overview by Dr. Dennis Gabor. New York: Optosonic Press,1970. Series title: State of the art review, no. 3.
    UCD Phys Sci Z7144.H6 K31970

Kaminow, Ivan P. *Laser devices and applications.* Ivan P. Kaminow and Anthony E. Siegman, eds. New York: IEEE Press,1973. Series title: *IEEE Press selected reprint series.*
    UCI Main Lib TA1684 .K36 Oversize
    UCLA Engr/Math TA 1675 K128 1973 Library
      has: Errata slip inserted

Kasper, Joseph Emil and Steven Feller. *The complete book of holograms :how they work and how to make them.* New York: Wiley, 1987. Series title: The Wiley science editions.
    UCD Phys Sci TAl540 .K37 1987
    UCLA Engr/Math TA 1540 K371987

____ . The hologram book. Englewood Cliffs, NJ.: Pren tice-Hall, 1985.
    UCI PhySciLib QC449 .K371985
    UCSB Library QC449 .K37 1985 Sci-Engrg
    UCSC Science QC449 .K371985
    UCSD Undergrad QC449 .K371985

Klein, H. Arthur. Holography. With an introd. to the optics of diffraction, interference, and phase differences 1st ed. Philadelphia: Lippincott, 1970. Series title: Introducing modern science.
    UCI PhySciLib QC449 .K57

Kock, Winston E. *Engineering applications of lasers and holography.* New York: Plenum Press, 1975. Series title: optical physics and engineering.
    UCD Phys Sci QC449 .K6 1975
    UCI PhySciLib QC449 .K6 1975
    UCR Phy Sci QC449 .K6 1975
    UCSB Library QC449 .K6 1975
    UCSC Science QC449.K61975
    CSL Main Lib QC119 K61975 Genl Coll

____ . *Lasers and holography*; an introduction to coherent optics. 1st ed. Garden City, N.Y: Doubleday, 1969. Series title: Science study series; [S62].
    UCD Phys Sci QC449 .K6
    UCI PhySciLib QC449 .K6
    UCSB Library QC449 .K6 Sci-Engrg
    UCSD S & E QC449 .K6
    CSL Main Lib QC449 K6 General ColI

___ . Lasers & holography: an introduction to coherent optics. 2nd enl. ed. New York: Dover Publications, 1981.
> UCB Physics QC449.K61981
> UO PhySciLib QC449 .K6 1981
> UCR Phy Sci QC449 K6x, 1981
> UCSD S & E QC449 .K6 1981

___ . Radar, sonar, and holography: an introduction. New York: Academic Press, 1973.
> UCD Phys Sci TK6575.K621
> UCI Main Lib TK6575.K62
> UCSB Library TK6575 .K62 Sci-Engrg
> UCSD S & E TK6575 .K62
> UCSD Scripps TK6575 .K7

Kurtz, Maurice K. "Potential uses of holography in photogrammetric mapping". 1971.
> UCSB Library TA1542 .K87 1971a Sci-Engrg
> Micro- film

___ . Study of potential application of holographic techniques to mapping; final technical report. Lafayette, IN: Purdue Research Foundation, Purdue University, 1971.
> UCR Rivera QC449.K87

Landry, Caliste John. "Ultrasonic imaging by Brillouin-Bragg diffraction: development of an operational system with prospective applications in medical diagnosis and material testing". 1972.
> UCSB Library TA1550 .L36 Sci-Engrg

Laser recording and information handling technology. Proceedings of a seminar held August 21-22, 1974, San Diego, CA. Leo Beiser, ed. Palos Verdes Estates, CA: Society of Photo-Optical Instrumen tation Engineers, 1975. Series title: Society of Photo-optical Instrumentation Engineers Seminar proceedings; 53.
> UCSD S & E TA1673.L37

Lasers and holographic data processing. N.C. Basov, ed. Translated from the Russian by P.S. Ivanov. Moscow: Mir Publishers, 1984. Series title: Advances in science and technology in the
> USSR. Technology series.
> UCD Phys Sci TA1684.L341984
> UCI Main Lib TA1684 .L34 1984
> UCSD Central TA1684.L341984
> UCSD S&E TA1684.L34198

Lavigne, Richard C. *Assessment of changeable message sign technology.* McLean, VA: U.S. Dept. of Transportation, Federal Highway Administration, Research, Development, and Technology,1986.
> UCB Trans PB87-154431 Microfiche

Lee, Hua. "Development and analysis of the back-projection method for acoustical imaging". 1980.
> UCSB Library QC244.5 .L43 Sci-Engrg

Lehmann, Matt. *Holography; technique and practice.* London: Focal Press, 1970. Series title: The Focallibrary.
> LBL Bldg 50 QC449 .L523 1970
> UCI PhySciLib QC449 .L44
> UCSB Library QC449 .L44 Sci-Engrg
> UCSC Science QC449.L44
> UCSD S & E QC449 .L44
> UCSD Scripps QC449.L44

Light and its uses: making and using lasers, holograms, interjerometers,and instruments of dispersion readings from Scientific American. Introductions by Jearl Walker. San Francisco: W. H. Freeman, 1980.
> UCB Moffitt TA1688.L53
> UCB Physics TA1688.L53
> UCD Phys Sci TA1688.L53
> UCI Main Lib TA1688 .L53 Oversize
> UClA Physics TA 1688 L53 1980
> UCR Phy Sci TA1688.L531980
> UCSB Library TA1688 .L53 Sci-Engrg
> UCSC Science TA1688.L53
> U CSD Cen tral TA1688 .L53
> UCSD Undergrad TA1688 .L53

Light vistas light visions. Sponsored by the Department of Art, St. Mary's College. Notre Dame, IN: St. Mary's College, 1983.

UCLA Art N 6494 L3 L53 1983 Restricted Cage

Heinecken Collection Lingenfelder, P. G. *Holography manual; a compilation of laboratory techniques commonly used in the construction of holograms including refinements developed at NELG.*. San Diego, CA: Naval Electronics Laboratory Center, 1969. Series title: NELC Technical document 47.

UCR Rivera QC449.L55
UCSC Science QC449.L56

Liu, Charles Yau-chi. "Some topics in holographic image formation".1974.

UCSD Central QC 3.6 .L59 Spec ColI Archives Diss
UCSD S & E QC 3.6 .L59

Lucie-Smith, Edward. *Art In The Seventies* .lthaca, NY: Cornell University Press, 1980. Contrib: R. Berkhout, H. Casdin-Silver, P.Claudius. Lyons, Harold. Lasers, quantum electronics, holography: part 1 : Introduction to lasers: Engineering 823.1 : a five-day short course, July 7-11, 1975 : lecture notes. Harold Lyons, coord. Los Angeles: University of California, University Extension, 1975.

UCLA Engr/Math TA 1675 L99 1975

____ . *Lasers, quantum electronics, holography: part I, Introduction to lasers: Engineering 823.1*, June 17- 21, 1974 : lecture notes. Harold Lyons, coord. Los Angeles: University of California, University Extension, 1974.

UCLA Engr/Math TA 1675 L99 1974

Matthews, Barbara Kubitz. *"Application of holographic methods to the analysis of flexural vibrations of annular sector plates"*. 1976. SRLF D 0000442699 Type EXPLAIN SRLF for borrowing information. McNair, Don. How to make holograms. 1st ed. Blue Ridge Summit, PA: Tab Books,1983.

UCSD Central TA1540.M361983

Mensa, Dean L."Techniques for microwave imaging". 1980.

UCSB Library TK6580 .M45 Sci-Engrg

"A Multi-frequency synthetic detecting holography with high depth resolution". Peking, China: The Research Group of Holography, Chinese Academy of Geological Sciences, [s.n.], 1976.

UCSD Scripps TA1542.M8

NATO Advanced Study Institute on Optical and Acoustical Holography,1971: Milan. Optical and acoustical holography; proceedings of the NATO Advanced Study Institute on optical and Acoustical Holography, Milan, Italy, May 24-June 4, 1971. Ezio Camatini,ed. New York: Plenum Press, 1972.

LBL Bldg 50 QC449 .N15 1971
UCD Phys Sci QC449.NI81971
UCI PhySciLib QC449 .N18
UCLA Engr/Math QC 449 N2141971
UCSB Library QC449 .N18 1971 Sci-En
UCSC Science QC449.N31971
UCSD S & E QC449 .N18 1971
UCSD Scripps QC449.NI81971

Neumann, Don Barker. *"The effect of scene motion on holography"*. Columbus, OH: s.n., 1967.

UCSB Library QC449 .N48 Sci-Engrg

Nondestructive holographic techniques for structures inspection. R. K. Erf ... [et al.]. Wright-Patterson Air Force Base, OH: Air Force Materials Laboratory, Air Force Systems Command, 1972.

UCD PhysSci TA417.2 .N681972mfll

Okoshi, Takanori. *Three-dimensional imaging techniques.* New York: Academic Press, 1976.

UCD Phys Sci TR780.038
UCI Main Lib TR780.038
UCSB Library TR780 .038 Sci-Engrg
UCSD S&E TR780.038

Optical Computing Symposium, Darien, Conn., 1972. Digest of papers presented at the 1972 one-day-indepth optical Computing Symposium, April 12, 1972 at the NOrDton School, Darien, Connecticut. Naval Underwater Systems Center and IEEE Computer Society, Eastern... [s.l.] Institute of Electrical Engineers, 1972.

    UCSD S & E TAl630.0671972

optics and photonics applied to three-dimensional imagery (Image 3-D):presented as part of the optics, Phototonics, and /conics Engineering Meeting (OPIEM), November 26-30, 1979, Strasbourg, France. Bellingham, WA: Society of Photo-Optical Instrumentation Engineers, 1980. Series title: Society of Photo-optical Instrumentation Engineers Seminar proceedings; v. 212.

    UCSD S & E TK7882.16067

optics in engineering measurement: 3-6 December 1985, Cannes, France. William F. Fagan, chair,ed. Organized by SPIE-the International Society for Optical Engin eerin g, ANRT --Association Nation ale de ... Bellingham, WA: SPIE-the International Society for Optical Engineering,1986. Series title: Proceedings of SPIE-the International Society for optical Engineering; v. 599.

    UCB Engin TA1677.0671 1986
    UCD Phys Sci TS5lO 's63 v.599
    UCI Main lib TS5lO .P632 v. 599
    UClA Engr/ Math TA 1555 0671985
    UCSB Library TA1555 .0671 1985 Sci/En
    UCSC Science QC370.5.S6v.599
    UCSD S & E TA1677.0671986

optics in entertainment :January 20-21, 1983, Los Angeles, California. Chris Outwater, chair,ed. Bellingham, WA: SPIE-the International Society for Optical Engineering,1983. Series title: Proceedings of SPIE-the International Society for optical Engineering; v. 391.

    UCB Engin T357.061983
    UCD Phys Sci TS5lO,S63 v.391
    UClA Engr/Math TA 16320671983

optics in entertainment :January 26-27, 1984, Los Angeles, CA. Chris Outwater, chair, ed. Bellingham, WA: SPIE-the International Society for Optical Engineering, 1984. Series title: Proceedings of SPIE-the International Society for optical Engineering; v. 462.

    UCB Engin TAl540 .06 1984
    UCD Phys Sci TS510 .S63 v.462
    UCI Main lib TS510 .P632 v. 462
    UCLA Engr/Math QC 350 S6 v,462
    UCSB Library TK8315 .068 1984 Sci-En
    UCSD S&E TK8315.o681984

Optics, Photonics, and Iconics Engineering Meeting,1979: Strasbourg, France. Optics and photonics applied to three-dimensional imagery (IMAGE 3-D): presented as part of the Optics, Photonics, and Iconics Engineering Meeting (OPIEM), November 26-30, 1979, Strasbourg, France. Bellingham, WA: Society of Photo- optical In-strumen tation Engineers,1980. Series title: *Society of Photo-optical Instrumentation Engineers Proceedings;* v. 212.

    UCB Engin TS510 's6v.212
    UCD Phys Sci TS510 .S63 v.212
    UClA Engr/Math TA 15420691979
    UCSB Library TA1542 .069 1979 Sci-En

optics today. John N. Howard, ed. New York, N.Y: American Institute of Physics, 1986. Series title: Readings from Physics today ; no. 3.

    UCD Phys Sci QC371 .06861986
    UClA Physics TA 1520 062 1986

Ostrowskii, IU. I. *Holography and its application.* Translated from the Russian by G. Leib. Moscow : Mir, 1977. UCSB Library TA1540 .08713 Sci-Engrg Outwater, Chris. and Eric Van Hamersveld. Guide to practical holography. Beverly Hills, CA: Pentangle Press,1974.

    UCLA AGSMgmt QC 449 094g
    UCSB Library QC449 .09 Sci-Engrg

Periodic structures, gratings, moire patterns, and diffraction phenomena: July 29-August 1, 1980, San Diego, CA. C.H. Chi, E.G. Loewen, C.L. O'Bryan III, eds. Bellingham, WA: Society of Photo-optical Instrumentation Engineers,1981. Series title: *Proceedings of the Society of Photo-optical Instrumentation Engineers*; v. 240.

    UCB Engin TS510.S6v.240
    UCD Phys Sci TS510.S6.3 v.240
    UCLA Engr/Math QC417P471980
    UCSB Library QC417 .P47 Sci-Engrg
    UCSD S&E QC417.P47

Pethick, J. *On holography and a way to make holograms.* Ontario: Belltower Enterprises, 1971.

    UCI PhySciLib QC449 .P4
    UCSD Central x79218 Closed Stacks

Photonics applied to nuclear physics, 2: proceedings; Strasbourg, Council of Europe, 5-7 December 1984. Geneva: CERN, 1985. Series title: Nucleophot.

    LBL Bldg 50 TA1522 .E9 1984
    UCLA Physics QC 793 P461984

Photorefractive materials and their applications. P. Gunter, j.-P Huignard, eds. Contributions by A.M. Glass ... ret al.]. Berlin: Springer-Verlag, 1988. Series title: Topics in applied physics; v. 61, etc. LBL Bldg 50 T A1750 .P47 1988

    UCB Engin TA1750.P471988
    UCB Physics TA1750.P471988
    UCI Main Lib TA1750 .P47 1988
    UCLA Physics TA 1750 P47 1988
    UCR Phy Sci TA1750 .P47 1988
    UCSB Library TA1750 .P47 1988 Sci-En v.1
    UCSC Science TA1750 .P471988 Lib v.1.
    UCSD S&E TA1750.P471988

Pietsch, Paul. ShuffZebrain. Boston: Houghton Mifflin, 1981.

    UCB Moffitt QP406 .P53
    UCD Main Lib QP406 .P53
    UEI Biomed WL 103 P626s 1981
    UCLA Biomed WL 102 P624s 1981
    UCLA Ed/Psych QP 406 P53
    UCR Bio-Ag QP406 P53x,1981
    UCSD Central QP406.P53
    UCSF General QP406.P531981

Pisa, Edward J., S. Spinak & A.F.Metherell. *Color acoustical holography.* Huntington Beach, CA: Douglas Advanced Research Laboratories, 1969. Series title: Douglas Advanced Research Laboratories. Research communication 109.

    UCD Phys Sci QC244.5 .P56

Powers,John Patrick. *"Some aspects of the application of Bragg diffraction of laser light to the imaging and probing of acoustic fields".* 1970.

    UCSB Library T A1550 .P68 Sci-Engrg

Practical holography: 21-22January 1986, Los Angeles, CA. Tung H. Jeong, Jacques E. Lud,man chair,eds. Presented in cooperation with American Association of Physicists in Medicine ... ret al.] . Bellingham, WA: SPIE-The International Society for Optical Engineering, 1986. Series title: *Proceedings of SPIE-the International Society for Optical Engineering;* v.615.

    UCB Engin TA1540 .p731 1986
    UCD Phys Sci TS510 .s63 v.615
    UCI Main Lib TS510 .P632 v. 615
    UCLA Engr/Math TA 1540 P731986
    UCSB Library TA1540 .p73 1986 Sci-En
    UCSC Science QC370.5.s6 v.615
    UCSD S & E T A1540 .p73 1986

Practical holography II: 13-14 January 1987, Los Angeles, CA. Tung H. Jeong, chair/ed. Sponsored by SPIE-the International Society for Optical Engineering, in cooperation with Center for.. Bellingham, WA: SPIE-the International Society for Optical Engineering,1987. Series title: Proceedings of SPIE-the International Society for Optical Engineering; v. 747.

    UCB Engin TAI540 .P62 1987
    UCLA Engr/Math TA 1540 P741987
    UCSB Library TA1540 .P7341987 Sci-En
    UCSC Science QC370.5.S6v.747
    UCSD S & E TAI540 .P73 1987

Proceedings of the information processing and holography symposium ICALEO 83. Symposium heads: David Casasent, Milton T. Chang. Organized with American Society of Metals ... ret al.] . Sponsored by Laser Institute ... Toledo, OH: The Institute, 1984. Series title: UA (Series) ; v. 41.

    UCSB Library QC449 .P76 1984 Sci-En

Proceedings of the Inspection, Measurment [sic} and Control and Laser Diagnostics and Photochemistry, ICALEO '84. Donald Sweeney, Robert Lucht, eds. Organized in cooperation with ... The American ... Toledo, OH: LIA-Laser Institute of America, 1985. Series title: LIA (Series) ; v . 45, 47.
  UCSB Library TK7871 .3 .P76 1985 Sci-En

Processing and display of three-dimensional data: August 26-27, 1982, San Diego, CA. James J. Pearson, chair,ed. Bellingham, WA: SPIE-The International Society for Optical Engineering,1983. Series title:Proceedingr of SPJE-the International Society for Optical Engineering; v. 367.
  UCB Engin TK7882.16 P7
  UCD Phys Sci TS510.863v.367
  UCI Main lib TS510 .P632 v. 367
  UCLA Engr/ Math TK 7882 16 P741982
  UCSB library TK7882.16 P725 1983 Sc-En
  UCSD S & E TK7882.16 P863 1982

Processing and display of three-dimensional data II : August 23-24, 1984, San Diego, CA. James J. Pearson, chair,ed. Cooperating organizations, Optical Sciences Center, University of Arizona, ... Bellingham, WA: SPIE-the International Society for Optical Engineering, 1984. Series title: Proceedings of SPIE-the International Society for optical Engineering; v. 507.
  UCB Engin TK7882.16P71984
  UCD Phys Sci TS510.863 v.507
  UCI Main lib TS510 .P632 v. 507
  UCLA Engr/ Math TK 7882 16 P741984
  UCSB Library TK7882.16 P75 1984 Sci-En
  UCSD S & E TK7882.16 P7 1984

Progress in holographic applications: 5-6 December 1985, Cannes, France. Jean Ebbeni, chair, ed. Organized by SPIE-the International Society for Optical Engineering, ANRT -Association Nationale de la ... Bellingham, WA: SPIE-the International Society for Optical Engineering,1986. Series title: Proceedings of SPIE-the International Society for optical Engineering; v. 600.
  UCB Engin TA1540.P761 1985
  UCD Phys Sci TS510.863 v.600
  UCI Main lib TS510 .P632v. 600
  UClA Engr/Math TA 1540 P771985
  UCSB library TA1540 .P77 1986 Sci-En
  UCSC Science QC370.5.S6 v.600
  UCSD S&E TA1540.P761985

Progress in holography : 31 March-2 April 1987, The Hague, The Netherlands. Jean Ebbeni, chair/ed. Organized by ANRT-Association nationale de la recherche technique, SPIE-The International Society for ... Bellingham,WA: SPIE, 1987. Series title: Proceedings of SPIE-The International Society for Optical Engineering; v. 812.
  UCB Engin QC449 .P711987
  UCD Phys Sci TS510 .S63 v.812
  UCI Main lib TS510 .P632 v. 812
  UCLA Engr/Math TA 1540 p7741987
  UCSC Science QC370.5.S6v.812
  UCSD S & E QC449.P761987

"Project Search". Subcommittee on Feasibility of Au tomated Fingerprint Identification/Verification. An experiment to determine the feasibility of holograph assistance to fingerprint iden-tification. Sacramento, CA: 1972. Series title: Project Search Technical report, no. 6.
  UCD Law lib HV6074 .P743

SRLF D 0000234864 Type EXPLAIN SRLF for borrowing information. Ramos, George Urban. "I. On the fast fourier transform; II. On the computations in digital holography". 1970.
  UCI PhyScilib QA404 .R3

Recent advances in holography: February 4-5, 1980, Los Angeles, CA. Tzuo-Chang Lee, Poohsan N. Tamura, eds. Bellingham, WA: Society of Photooptical Instrumentation Engineers, 1980. Series title: Society of Photooptical Instrumentation Engineers Proceedings; v. 215.
  UCB Engin TS510.S6v.215
  UCD Phys Sci TS510 .S63 v.215
  UCIA Engr/Math TA 1542 R42 1980
  UCSB library TA1542 .R42 Sci-Engrg
  UCSD S & E TA1540 .R43

Saxby, Graham. Practical holography. New York, N.Y.: Prentice-Hall International, 1988.
  UCB Engin TA1540.83611988
  UCD Phys Sci TA1540.S361988
  UCI Main lib TA1540 .S36 1988
  UClA Engr/Math TA 1540 S361988
  UCR Phy Sci TA1540.S361988

Schlussler, Larry. "Improvement of the horizontal resolution of a Bragg-diffraction imaging system and motion limitations of a holographic system". 1978.
    UCSB Library TA1550 .S34 Sci-Engrg

Schueler, Carl Frederick. "Development and applications of computer-assisted acoustic holography". 1980.
    UCSB Library TA1550 .S38 Sci-Engrg

Schumann, Walter and J.-P. Zurcher, D.Cuche. Holography and deformation analysis. Berlin: Springer-Verlag, 1985. Series title: Springer series in optical sciences; v. 46.
    UCB Physics TA1542 .S381 1985
    UCD Phys Sci QC450.S65 v.46
    UCI Main Lib TA1542 .s38 1985
    UCLA Engr/Math TA 1542 S381985
    UCSB Library TA1542 .S38 1985 Sci-En
    UCSC Science TAI542.S381985
    UCSD S & E TAI542 .S38 1985

Schwank, James Ralph. "Refractive holography". 1974.
    UCLA Engr/Math LD 791.8 E4174 S398 Micro film

Sherman, George Charles. "Wavefront reconstruction and its application to the study of the optical properties of atmospheric aerosols". 1969. SRLF D 0000264507 Type EXPlAIN SRLF for borrowing information. Shuman, Curtis Alan. "Holographic imaging through moving diffusive media". 1973.
    UCSD Central QC 3.6 .S566 Spec ColI Archives Diss
    UCSD S & E QC 3.6 .S566

___ " "Holographic imaging through moving scatterers" .1972.
    UCSD Central QC 3;7 .s4 Spec Coll Archives Diss
    UCSD S&E QC3.7.S4

Smith, Howard Michael. Principles of holography. New York, Wiley-Interscience, 1969.
    LBL Donner QC449 .S649 1969
    UCI PhySciLib QC449.S6
    UCLA Engr/Math QC 449 S649p
    UCR Phy Sci QC449.S61969
    UCSB Library QC449 .S6 Sci-Engrg
    UCSC Science QC449.S6
    UCSD S & E QC449.S6
    CSL Main Lib QC449 S6 General Coll

___ " Principles of holography. 2d ed. New York: Wiley,1975.
    LBL Donner QC449 .S649 1975
    UCB Moffitt QC449.S6 1975
    UCD Phys Sci QC449.S61975
    UCI PhySciLib QC449 .S61975
    UCSB Library QC449.S6 1975 Sci-Engrg
    UCSC Science QC449.S61975
    CSL Main Lib QC449 S6 1975 Genl ColI

Solem, Johndale C. High-intensity X-ray holography: an approach to high-resolution snapshot imaging of biological specimens. Los Alamos, N.M.: Los Alamos National Laboratory, 1982.
    UCLA Biomed TA 1540 S685h 1982

Solymar, L. and DJ. Cooke. Volume holography and volume gratings. London: Academic Press, 1981.
    UCB Physics QC449.S61
    UCD Phys Sci QC449.S65 1981
    UeI PhySciLib QC449 .S661981
    UClA Physics QC 449 S641981
    UCR Phy Sci QC449 .S72 1981
    UCSB Library QC449.S64 1981 Sci-Eng
    UCSD S & E QC449 .S65 1981

Soroko, Lev Markovich. Holography and coherent Optics. Translated from Russian by Albin Tybulewicz; with a foreword by George W. Stroke. New York: Plenum Press, 1980.
    UCB Engin QC449.S6713
    UCB Physics QC449.S6713
    UCD Phys Sci QC449 .S6713
    UCI PhySciLib QC449 .S6713
    UCLA Physics QC 449 S671319S0
    UCSB Library QC449 .S6713 Sci-Engrg
    UCSD S & E QC449 .S6713

Stone, William Ross. "The concept, design, and operation of a demonstration holographic radio camera". 1978.
    UCSD S & E QC 3.6 .S765

_____ . "A remote probing technique for inhomogeneous media and an application to the study of satellite scintillations". 1974.
    UCSD Central QC 3.7 .S765 Spec Coll Archives
        Diss
    UCSD S & E QC 3.7 .S765

Strand, Timothy C. "Comparison of analog and binary holographic data storage" .1973.
    UCSD Central QC 3.7 .S77 Spec Coll Archives
        Diss
    UCSD S & E QC 3.7 .s77

Stroke, George W. An introduction to coherent optics and holography. New York: Academic Press, 1966.
    LBL Donner QC357 .S7 1966
    UCB Optom QC357.S96
    UCD Phys Sci QC357.S96
    UCI PhySciLib QC357 .S96
    UCLA Engr/Math QC 357 S919i
    UCR PhySci QC357.S86
    UCSB Library QC357 .S8 Sci-Engrg
    UCSC Science QC357.S96
    UCSD S&E QC357.S919
    UCSD Scripps QC357.S919

_____ . An introduction to coherent optics and holography. 2d ed. New York: Academic Press, 1969.
    LBL Bldg 50 QC357.S71969
    LBL Donner QC357.S71969
    UCB Moffitt QC357 .S96 1969
    UCB Optom QC357.S96 1969
    UCD Phys Sci QC357.S961969
    UCI PhySciLib QC357 .S961969
    UCLA Engr/Math QC 357 S919i 1969
    UCR PhySci QC357 .S861969
    UCSB Library QC357.S8 1969 Sci-Engrg
    UCSC Science QC357.S961969
    UCSD S&E QC357.S9191969
    UCSD Scripps QC357 .S919 1969
    UCSF General QC357 .S921i 1969

Su, Kung-Yen. "The fabrication of an optoacoustic transducer for real-time diagnostic imaging systems". 1982.
    UCSB Library TA1770 .S9 1982 Sci-En

Sutton, Jerry Lee. "Broadband acoustic imaging". 1974.
    UCSD Central QC 3.7 .S87 Spec Coll Archives
        Diss
    UCSD S & E QC 3.7 .S87

Symposium on Applications of Holography in Mechanics, 1971: University of Southern California. Symposium on Applications of Holography in Mechanics. W. G. Gottenberg, ed. New York: American Society of Mechanical Engineers, 1971.
    UCD Phys Sci TA349.S91971
    UCI Main Lib TA349 .S9
    UCLA Engr/Math TA 350 S987s 1971

Tarasov, L. V. Laser age in optics. Translated from the Russian by V. Kisin. Moscow: Mir Publishers, 1981.
    UCD Phys Sci QC446.2 .T3713 1981
    UClA Physics QC 446.2 T3713 1981
    UCSB Library QC446.2 .T3713 1981 S& E

Three-dimensional imaging; April 21-22, 1983, Geneva, Switzerland. Jean Ebbeni, Andre Monfils, chairmeneditors. Bellingham, WA: SPIE-the International Society for Optical Engineering,1983. Series title: Proceedings of SPIE-the International Society for Optical Engineering; v. 402.
    UCB Engin TK7882.I6 T47 1983
    UCD Phys Sci TSS10 .S63 v.402
    UCI Main Lib TSS10 .P632v. 402
    UCLA Engr/Math TK8315 T481983
    UCR PhySci TK7882.I6T471983
    UCSB Library TK7882.I6 T469 1983 S&En
    UCSD S&E TK7882.16T4741983

Three-dimensional imaging: August 25-26, 1977, San Diego, CA. Stephen A. Benton, ed. Presented by the Society of Photo-optical Instrumentation Engineers, in conjunction with the IEEE Computer ... Bellingham, WA: SPIE, 1977. Series title: Society of Photooptical Instrumentation Engineers Proceedings ; v. 20.

> UCB Engin TS510.S6v.120
> UCD Phys Sci TS510.863 v.120
> UClA Engr/Math TA 1632 T413 1977
> UCSB library TK7882.16 T47 Sci-Engrg
> UCSD S & E TK7882.16 T47

Topical Meeting on Halogram Interferometry and Speckle Metrology, 1980, June 2-4 : North Falmouth, MA. A di~st of technical papers presented at the Topical Meeting on Hologram Interferometry and Speckle Metrology, June 2-4, 1980, Sea Crest Hotel, North Falmouth, Cape Cod, MA. [s.1.]: Optical Society of America,1980.

> UCR Phy Sci QC449 .T674 1980
> UCSC Science QC449.T661980
> UClA Engr/Math TA 1555 T661980

Tricoles, Gus Peter. "Some topics in microwave holography". 1971.

> UCSD Central QC 3.6 .T743 Spec Coli Archives
>   Diss
> UCSD S & E Q!2 3.6 .T743

Trolinger, J. D. Laser instrumentation for flow Judd diagnostics. S. M. Bogdonoff, ed. Neuilly-sur-Seine, France: North Atlantic Treaty Organization, Advisory Group for Aerospace Research and Development, 1974. Series title: AGARDograph; no. 186.

> UCD Phys Sci TL500 .N6 no.1861974
> UClA Engr/Math TA 1522 T747l

Tse, Nie But. "Digital reconstruction of acoustic holograms". 1979.

> UCSB Library TA1550 .T73 Sci-Engrg

United States;Japan Science Cooperation Seminar on Pattern Information Processing in Ultrasonic Imaging,3rd: 1973 : University of Hawaii. Ultrasonic imaging and holography: medical, sonar, and optical applications: [proceedings]. George W. Stroke ... ret al.],ed. New York: Plenum Press, 1974.

> LBL Donner Q!2244.5.U581973
> UCD HealthSci QC244.5 U541973
> UCD Phys Sci R895.A1 U541973
> UCI Biomed W3UN611973u QC244.5
> U581973u
> UCI Main Lib R895A1 U541973
> UCR Phy Sci R895A1 U54 1973
> UCSB library R895A1 U541973 Sci-En
> UCSD Biomed QT 34 U6091973u

United States-Japan Seminar on Information Processing by Holography, 2nd :1969 : Washington, D.C. Applications of holography; proceedings. Euval S. Barrekette, ed. New York: Plenum Press, 1971.

> LBL Bldg 50 QC449.U581969
> UCD Phys Sci QC449 .U5 1969
> UCI PhyScilib QC449 .U5 1969
> UClA Engr/Math QC 449 U58a 1969
> UCR PhySci QC449.U51969
> UCSB Library QC449 .U5 1969 Sci-En
> UCSD S & E QC449 .U5 1969
> UCSD Scripps QC449 .U5 1969

Unterseher, Fred, Jeannene Hansen and Bob Schlesinger. Holography handbook: making holograms the easy way. Berkeley, CA: Ross Books, 1982.

> UCB Moffitt TAl542 .U57 1982
> UeI Main lib TA1542.U57
> UCSB library TA1542 .U571982 Sci-En
> UCSC Science TA1542.U57 1982

Vest, Charles M. Holographic interferometry. New York: Wiley, 1979. Series title: Wiley series in pure and applied optics.

> UCB Engin TA1555 .V47
> UCB Physics TA1555.V47
> UCD Phys Sci TA1555.V471
> UCI Main Lib TA1555.V47
> UClA Physics TA 1555 V6381979
> UCR Phy Sci TA1555 .v47 1979
> UCR Rivera TA1555.V471979
> UCSB library TA1555 .V47 Sci-Engrg
> UCSC Science TA1555.V47
> UCSD S & E TA1555.V47

Vourgourakis, Emmanuel John's. "Coherence limitations on holographic systems". 1967.
　　UClA Engr IMath LD 791.9 E4 V945

Microfilm Walton, Paul. Space-light: a holography and laser spectacular. London: Routledge & Kegan Paul, 1982.
　　UCD Main Lib N6494.L3 W31982
　　UCI Main Lib N6494.L3 W341982b
　　UCSB Arts Lib N6494.L3 W341982 Arts
　　UCSD Central N6494.L3 W341982b

Wang, Keith Yu-Chih. "Threshold contrasts for various acoustic imaging systems". 1972.
　　UCSB Library TA1550 .W36 Sci-Engrg

Wenyon, Michael. Understanding holography. New York: Arco Pub. Co., 1978.
　　UCB Moffitt TAl540.W46
　　UCR Phy Sci TA1540 .W46 1978
　　UCSC Science TA1540.W46

___ . Understanding holography. Newton Abbot, Eng.: David & Charles,1978.
　　UCSB Library TA1540 .W461978b S&En

___ .. Understanding holography. 2nd Arco ed. New York: Arco Pub., 1985.
　　UCD Phys Sci TAI540 .W46 1985
　　UCSD Central TA1540 .W46 1985

Wollman, Michael Thomas. "An experimental acoustical holographic system for eventual use in the ocean".1975.
　　UCSB Library TA1550 .W64 Sci-Engrg

Yu, Francis T.S. Introduction to diffraction, information processing, and holography. Cambridge, MA: MIT Press, 1973.
　　LBL Bldg 50 QC415.)941973
　　UCD Phys Sci QC415.Y8
　　UCI PhySciLib QC415 .Y8
　　UCR PhySci QC415.Y81973
　　UCSB Library QC415.Y8 Sci-Engrg
　　UCSC Science QC415.Y8
　　UCSD S & E QC415.Y8

International Conference on Holographic Systems, Components and Applications, 1987: Churchill College. Holographic systems, components and applications Churchill College, Cambridge, 10th-12th September, 1987. London: Institution of Electronic and Radio Engineers, 1987. Series title: Publication / Institution of Electronic and Radio Engineers; no. 76.
　　UCD Phys Sci TA1542 .H6 1987
　　UCSD S & E TAI542 .H643 1987

International Conference on Computer-generated Holography: 2nd: 1988: Los Angeles, CA. Computer-generated Holography II : 11-12 January 1988, Los Angeles, California, [proceedings] . Sing H. Lee, chair/ed. Sponsored by SPIE-The International Society for Optical Engineering ;... Bellingham, WA: SPIE-The International Society for Optical Engineering, 1988. Series title: Proceedings of SPIE-the International Society for Optical Engineering; v. 884.
　　UCSC Science QC370.5.S6 v. 884
　　UCSD S&E TAI542.1591988

Holographic optics: design and applications: 13-14 January 1988, Los Angeles, CA. Ivan Cindrich, Chair/ Ed. Sponsored by SPIE-The International Society for Optical Engineering ; cooperating ... Bellingham, WA: SPIE-The International Society for Optical Engineering, 1988. Series title: Proceedings of SPIE-The International Society for Optical Engineering; v. 883.
　　UCSC Science QC370.5.S6 v.883

PERIODICALS

Acoustical Holography. International Symposium on Acoustical Holography. New York: Plenum Press. v.I-7, (1969-1977).
　　UCB ENG! QC244.5.A1.A25 bound v.I-7(1969-
　　　　1976)
　　UCD PHYSCI QC244.5 A36 v.I-7 (1969-1976)
　　UClA CLU-EMS QC 244.5 Al 161 1973-1976
　　UCR PHYSCI QC244.5.151 b.v.5-7 (1973-76)
　　UCSB CU-SB QC 244.5.15 v.5-7,1973-76
　　UCSC SCI UB QC244.5.15 v.5-7(1973-76); in
　　　　McHenry Library

UCSD S&E QC 244.5 A25 Stacks v.5-71973-76
UCSD SCRIPPS SIO 1AC3664 Fl.2v .. 1-7,
1969-1976

Acoustical holography, proceedings. International
Symposium on Acoustical Holography and Imaging.
(-1973).
   UCD l-ll..1H SCI QC244.5I5 v.4,1972
   UCSD S & E QC 244.5 A25 Stacks v.l-4 (1967-72)
   UCI PSL QC 244.5 15 v. bl-4(1967-72)
   UCR PHY SCI QC244.5.15 b.v.1-4(1967-72)
   UCSB CU-SB QC 244.5. 15 v.l-4, (1967-72)
   UCSC SCI LIB QC244.5.15 v.l(1967-72); In
      McHenry Library.
   UCSF GEN'L QC244.5.I5 1969-72 v.1, 437137
      (1969);v.4,437138(1972)

Acoustical imaging and holography. New York:
Crane, Russak,1978-1979.
   UCB ENGI QC244.5.Al.A26 bound 1:1
      (1978)-1:2(1979)
   UCSB CU-SB QC 244.5. A26 v.1, 1978-79
      (ceased pub!.); ltd loan

Advances in holography. New York, M. Dekker,
1975-76.
   UCB ENGI QC449.A381 bound 1-3(1975-76)
   UCD PHY SCI QC449 A38 v.1-3, (1975-76)
   UeI PSL QC 449 A38 v.bl-3, (1975-76)
   UCSB CU-SB QC 449. A38 v.I-3, (1975-76)
      (ceased publication)
   UCSC SCI LIB QC449.A38 v.1-3(1975-76)
   UCSD S & E QC 449 A3 Stacks v.I-3 (1975- 76)

Acoustical holography; [proceedingsj:Acoustical im-
aging, 1978. International Symposium on Acoustical
Holography and Imaging. New York: Plenum Press,
1978.
   UCI PSL QC 244515 b5-7(1973-76)
   UCSF GENERAL QC244.5.I5 1974-1977
   UCSB CU-SB QC 244.5.15 v.8, 1978
   UCSC SCI LIB QC244.5.15 v.8(1978); in
      McHenry Library.
   UCSDS&E
   UCI PSL QC 244.5 A25 Stacks v.8 (1978) v.5-7
   QC 244.5 15 b.v.8(1978)

Afterimage. Rochester, NY: Visual Studies Workshop,
V.12, no.7, Feb. 1985. See: A. Sargent-Wooster, "Man-
hattan shortcuts, Harriet Casdin-Silver's 'Thresholds'"
p 19. Fundamentals and applications of optical data
processing and holography. Ann Arbor: University
Michigan Engineering Summer Conferences. Hologra-
phy.
   UCB CHEM QC449.A12K25 1970
   UCB ENGI QC449.A12K25 1969; 1971/72
   U CSB CU-SB Z 7144. H6 H65 1969-72
   UCSD S & E QC 1 H77 Stacks 1969-1972

Holography News. (newsletter). Washington, DC.:
Louis Kontnick, since 1987. (see listing for address).
The Holo-gram. (newsletter). Allentown, PA: Frank
DeFreitas, since 1983. (See listing for address). New
Scientist. London, England: 1977. See: R. Weale
"Art:Holography by H.Casdin-Silver" (June29).

Titles in French

Caussignac, Jean Marie. Visualisation d'ecoulements aerodynamiques dans les compresseurs par interferometne holographique. Chatillon, France: Office national d'etudes et de recherches aerospatiales, 1972. Series title: France. Office national d'etudes et de recherches aerospatiales Note technique, 190.
UCSD S&E TI.1.F7852v.190

Francon, M. Holographie. Paris: Masson, 1969. Series title: Recherche appliquee.
UCSB library QC449 .F7 Sci-Engrg

International Symposium of Holography, Besancon, France, 1970. Applications de l'holographie; comptes rendus du Symposium international d'holographie. Applications of holography; proceedings of the International Symposium of Holography. Besancon 6-11 juillet 1970. Besancon: Laboratoire de physique generale et optique, Universite de Besancon,1970.
UCD Main lib QC449 .1571970
UCSB Library QC449 .157 1970 Sci-Engrg
UCSD S & E QC449.1571970

Pinson, G., A. Demailly, and D. Favre.La Pensee: approche holographique. Lyon: Presses universitaires de Lyon, 1985. Series title: Collection Science des systemes. Serie theone des systemes.
UClA URL TA 1542 P56 1985

Titles in German

Claus,Jurgen. ChippppKunst : Computer, Holographie, Kybernetik, Laser. Originalausg. Frankfurt/M.: Ullstein, 1985. Series title: Ullstein Materialien.
UCSB library TK7885 .C551985 SciEngrg

Laserbeugung an elektronenmikroskopischen Aufnahmen. Ludwig Reimer ... [et al.]. Opladen: Westdeutscher Verlag, 1973. Series title: Forschungsberichte des Landes Nordrhein-Westfalen ; Nr. 2314.
UCI Main lib QH212.E4 L37
UCSB library T3 .F65 no.2314 SciEngrg

Licht-Blicke : Holographie, die 3. Dimension fur Technik und kunst : [Ausstellung] 7. Juni-30. September 1984, Deutsches Filmmuseum Frankfurt am Main. Schirmherr, Bundesprasident a. D. Walter Scheel; ... Frankfurt am Maain : [Deutsches Filmmuseum],1984. Interviews with: S.Benton, M.Benyon, R.Berkhout, H.Casdin-Silver, F.Mazzero, S.Moree, N.Phillips, G. Schneider-Siemssen, D.Schweitzer. Articles by: M. Schneckenburger, et alia.
UCl PhyScilib QC449 .L521984
UCR Rivera fN6494.L3 L5 1984 Cage

Mehr Licht: Kunstlerhologramme und Lichtobjekte = More light: [artists's [sic] holograms and light objects]. herausgegeben von Achim Lipp und Peter Zec. Hamburg: Fielmann im E. Kabel Verlag, 1985.
UCB Main N6494.L3 M4571 1985
UCR Rivera N6494.L3 M4571985 Cage
UCSC McHenry N6494.L3 M457 1985
UCSD Central N6494.L3 M457 1985

Menzel, Eric, W. Mirande [und] I.Weingartner. Founer-Optik und Holographie . Wien: Springer-Verlag, 1973.
UCSB Library QC403 .M4 Sci-Engrg

Optoelectronik in der Technik : Vortrage des 6. Internationalen Kongresses Laser 83 optoelektronik = Optoelectronics in engineering : proceedings of the 6th International Congress, Laser 83 Optoelektronik / herausgegeben. Berlin: Springer-Verlag, 1984.
UCSD S & E ~73 .06751983

Schreier, Dietmar unter Mitarbeit von W. Hase .. [et aLl Synthetische Holografie. Weinheim: PhysikVerlag, 1984.
UCI Main lib TA1540 .s37 1984

Universitatsbibliothek Jena. Zusammenstellung inund auslandischer Patentschriften auf dem Gebiet tier Holographie. Berichtszeit: 1948-1970. Gesamtleitung: Konrad Marwinski, Informationsabt., 1971 . Series title: Universitatsbibliothelt Jena Bibliographische Mitteilungen, Nr. 12.
VCB Main QC449A12 V541 1971
ven Phys Sci Z7144.H6 J4

Zec, Peter. Holographie: Geschichte, Technilt, Kunst. Koln: DuMont,1987.
VCSB Arts lib N6494.L3 Z42 1987 Arts

Titles in Swedish

Holografi: det 3-dimensionella mediet. New York: Museum of Holography; Stockholm: distribution, AVC, 1976.
VCR Rivera fTA1542
H62 Cage

Titles in Russian

Bakhrakh, L.D. i S.D. Kremenetskii. Metody izmerenii parametrov iizluchaiushchikh sistem v blizhnei zone. Leningrad: Izd-vo "Nauka", Leningradskoe otdelenie, 1985.
VCB Engin QC449 .M4 1985

Bakhrakh, L.D i V.A. Makeeva. Primenenie golografii v meditsine i biologii. Leningrad: Nauka, 1977.
VCB Biology R857.H64 p75 1977

CRL GenCollec B-37831 Type EXPlAIN CRL for borrowing information. Barachevskii, V.A.Svoistva svetochuvstvitel'nykh materialov i ikh primenenie v golografii: sbornik nauchnykh trudov. Otvetstvennyi redaktor Leningrad: Izd-vo "Nauka," Leningradskoe otd-nie, 1987.
VCB Engin TAl540.S8611987

Barachevskii, V.A. Neserebrianye i neollychnye sredy dlia golografii. Leningrad: "Nauka", Leningradskoe otd-ie, 1978.
VCB Physics QC449.N41978

Denisiuk, IV.N. Opticheskaia golografiia : prakticheskie primeneniia .. Leningrad: Izd-vo "Nauka," Leningradskoe otd-nie, 1985.
CRL GenCollec B-37116 Type EXPlAIN CRL for borrowing information.

___ .. Opticheskaia golografiia: [Sb. statei}. AN SSSR, Fiz.-tekhn. in-t im. A.F. Ioffe, Nauch. sovet po probl. "Golografiia". Leningrad: Nauka, Leningr. otd-nie, 1979.
VCB Main TA1542 .068

Derkach, M.F.Dinamicheskie spektry rechevykh signalov. L'vov: Izd-vo pri L'vovskom Gos. universitete Izdatel'skogo ob"edineniia "Vyshcha shkola", 1983.
UCB Main PG2135.D471983
UCD Main Lib PG2135.D4 1983
UClA URL PC 2135 D55 1983
UCSB Library PG2135.D471983

Gurevicha, S.B. i V.K. Sokolova.Primenenie metodov opticheskoi obrabotki informatsii i golografii. Leningrad: LIIAF, 1980.
UCB Engin TA1630.P741 1980
CRL GenCollec B-33691 Type EXPLAIN CRL for borrowing information.

Gurevicha, S.B., O.A. Potapova."Golografiia i opticheskaia obrabotka informatsii v geologii". Dokl. seminara. Leningrad: Akademiia Nauk SSSR, Fiziko-tekhnicheskii in-t im. A.F. Ioffe, 1980.
NRLF $B 773 346 Type EXPLAIN NRLF for borrowing information.

IAkovkin, I. B. Difraktsia sveta na akusticheskikh poverkhnostnykh volnakh . otv. redaktor S.V. Bogdanov. Novosibirsk: Izd-vo "Nauka", Sibirskoe otd-ie, 1979.
UCB Physics QC176.8A3 121979

IAroslavskii, L. P. and N.S. Merzliakov. TSifrovaia golografiia. Moskva: Izd-vo "Nauka", 1982.
UCB Engin TA1542 .1281 1982
CRL GenCollec B-34155 Type EXPLAIN CRL for borrowing information.

International School on Coherent Optics and Holography,2nd: 1981: Varna, Bulgaria. Integral'naia optika, volokonnaia optika i golografiia : materialy vtoroi Mezhdunarodnoi shkoly po kogerentnoi optike i golografiiVarna '81,28.09-03.10.1981, Varna, Bolgariia. Redaktsionnaia kollegiia P. Simova, ... Sofiia: Izd-vo Bolgarskoi akademii nauk, 1982.
UCB Physics TAI660 .1565 1981
UCD Phys Sci TAI660 .15651981
CRL GenCollec A-39257 Type EXPLAIN CRL for borrowing information.

Kirillov, N. I. Vysokorazreshaiushchie fotomaterialy dlia golografii i protsessy ikh obrabotki. Moskva: Nauka, 1979.
UCB Engin TAI542.K57

Klimenko, I. S. Golografiia sfokusirovannykh izobrazhenii i spekl-interferometriia. Moskva: "Nauka," Glav. red. fiziko-matematicheskoi lit-rio 1985.
CRL GenCollec B-36771 Type EXPLAIN CRL for borrowing information.

Kulakov, Sergei Viktorovich. Akustoopticheskie ustroistva spektral'nogo i korreliatsionnogo analiza signalov. Akademiia nauk SSSR, Nauchnyi sovet po probleme" Golografiia", Fiziko-tekhnicheskii institut imeni A.F. Ioffe. Leningrad :"Nauka", Leningradskoe otd-nie, 1978.
UCB Engin TA1770.K8

Petrashen, G. I. Prodolzhenie volnovykh polei v zadachakhs seismorazvedki. Leningrad: "Nauka," Leningr. otd-nie, 1973.
UCLA Geology QE541 P471973

Radiogolografiia i opticheskaia obrabotka informatsii v mikrovolnovoi tekhnike : [Sbornik statei}. Akademiia nauk SSSR, Otdelenie obshchei fiziki i astronomii, Nauchnyi sovet po probleme "Golografiia" Leningrad: "Nauka," Leningradskoe otd-nie, 1980.
UCB Engin QC675.8.R31 1980

Sobol eva, G.A. R.egistriruiushchie sredy dlia izobrazitelnoi golografii i kinogolografii : [Sb. statei} . AN SSSR, Otd-nie obshch. fiziki i astronomii, Nauch. sovet po probl. Golografiia. Leningrad : Nauka, Leningr. otd-nie, 1979.
UCB Engin TA1542 .R44

Sokolov, A. V. i IA.A. Al'tmana. Primenenie metodov optichesltoi golografii dlia issledovaniia biologicheskikh mikroob"ektov. Leningrad: Nauka, Leningradskoe otd-nie ,1978. Series ti tIe: Metody Jiziologicheskikh issledovanii.
UCB Biology QC449.S611978

Voropaev, N. D. Anglo-russ1r.ii slovar' po 1r.vantovoi elektronike i golografii : Okolo 18000 terminov . Pod red. A. M. Leontovicha. Moskva: Rus.iaz., 1977.
    UCR Phy Sci QC687 .V671977 Ref
    UCR Phy Sci QC5 .V58 Ref

Vsesoiuznaia shkola po golografii, 6th: 1974: Yerevan, Armenian S.S.R. Materialy VI Vsesoiuznoi shkoly po golografii : 11-17 feuralia 1974 g .. redaktory, G.V. Skrotskii, B.G. Turukhano, N. Turukhano]. Leningrad: LIIAF, 1974.
    CRL GenCollec B-37162 Type EXPLAIN CRL for borrowing information.

Vsesoiuznaia shkola po golografll, 7th : 1975 : Rostov, R.S.F.S.R. Materialy VII Vsesoiuznoi shkoly po golografii : ianvar' 1975 g. [podgotovleny k pechatki N. Turukhano]. Leningrad: Leningradskii in-t iadernoi fiziki, 1975.
    NRLFB32195

Ze1'dovich, B. lA, N.F. Pilipetskii, & V.V.Shkunov. Obrashchenie volnovogo fronta. Moskva: "Nauka," Glav. red. fiziko-matematicheskoi lit-ry, 1985.
    UCB Engin QC446.2.Z4411985
    CRL GenCollec B-37082

EXHIBIT CATALOGUES & MISCELLANY

18a. Bienal Internacional de Sao Paulo, Catalogo Geral. Sao Paulo, Brasil: Fundacao Bienal de Sao Paulo, 1985. See "Entre a Ciencia EA Ficcao", pp167-197, re: holographers M.Baumstein, H. CasdinSilver, J.W.Garcia.

Alice in the Light World. Tokyo, Japan: The Ashai Shimbun, 1978.

ARTTRANSITION. Cambridge, MA: MIT Center for Advaced Visual Studies/University Film Study Center, 1975. See H. Casdin-Silver "Holography... " pp30-32.

Critic's Choice: The Craft of Art; Peter Moore's Liverpool Project 5. Liverpool, England: November 3, 1979. See E. Lucie-Smith, "New Attitudes, New Materials, New Techniques".

Electra 83. Paris: Les Amis du Musee d'Art de la Ville de Paris, 1983. See F.Popper, "Electricity and Electronics in the Art of the 20th century", pp 46-50.

EXPANSION. Internationale Biennale fur Graphik und Visuelle Kunst. Horst Gerhard Haberl, Generalsekretar. Wien, Austria: Internationale Biennale fur Graphik und Visuelle Kunst, 1979. See O.Peine "MIT-Center for Advanced Visual Studies", "Sky Events" pp 232-239; E. Goldring "Documentation room", "Centerbeam"p 232; H.Casdin-Silver "Holography, a holographic environment" p 234; G.Kepes "Art of the Environment" p 99. Fantasy of Holography. Tokyo, Japan: Seibu Museum of Art, 1976. Itsuo Sakane, ed. Contr: Shuntoro Tankawa, Junpei Tsujiuchi.

Harriet Casdin-Silver Holography. New York: Museum of Holography, 1977. First one person exhibition at the Museum.

High Technology and Art 1986. Tokyo, Japan: Tokyo Shimbun & Nagoya Shimbun, with Associates of Art and Technology, Japan.

'Holography redefined', 'Thresholds'. Harriet Casdin-Silverwith Dov. Eylath. New York: Museum of Holography, 1984. Group exhibition.

Inter. Quebec, Canada: Les Editons Intervention, Printemps 86, no.31, 1986. See E. Shapiro, "Art, Perceptionet Holographie" p. 32; L.Heaton, "Tire D'une Entrevue avec Harriet Casdin-Silver" pp32-3.

Images in Time and Space. Ottawa, Ontario, Canada: Association of Science and Technology Inc, 1987. Travelling exhibit.

International Holography. London, England: The Photographers' Gallery, 1980.

Light and Substance. New Mexico: University of New Mexico Art Museum, 1973-75. History of photography, holography by S. Benton, H.Casdin-Silver. Van deren Coke, org.

MultiMedia Exhibition. Kansas City, MO: Nelson Gallery at Atkins Museum of Fine Arts, 1970. See "Holography by H.Casdin-Silver".

Otto Peine und CAYS: 20th Anniversary CA VS. Karlsruhe, West Germany: Badischer Kunstverein, 1988. See H . Casdin-Silver, A. Cheji, D. Jung, J .Powell.

Sky art conference '83. Cambridge, MA: MIT Center for Advanced Visual Studies with der Landeshaupstadt Munchen der BMW AG und der Digital Equipment GmbH, 1983

CHAPTER 9

GLOSSARY

Absorption Hologram: A hologram formed in a material which acquires a certain density in response to exposure. When the hologram is illuminated, part of the light which is not absorbed is diffracted into forming the image.

Achromatic: Black and white. In optical systems, the term is used to describe lenses which correct for chromatic aberration.

Acoustical Holography: The making of holograms by using sound waves.

Additive Color Mixing: Means by which two or more frequencies are combined by superimposition to create more colors.

Ambient Light: Light present in the immediate environment. In holographic display, often used to describe background light that is not part of the hologram illumination and may interfere with the viewing of the image.

Amplitude: The maximum value of the displacement of a point on a wave front from its mean value. Graphically, the height or depth of the crest or trough of a wave from its zero point.

Amplitude Hologram: A hologram by which information is stored as variations in transmittance. Also called absorption hologram.

Antihalation Backing (AH): A dark material placed on the back surface of a plate or film to prevent unwanted light from striking the emulsion. Helps prevent the formation of "Newton Rings" in the hologram. Only to be used with transmission holograms.

Argon Laser: A laser which operates when argon gas is ionized and controlled by a magnetic field. Produces several blue and green frequencies.

Astigmatism: An aberration caused by the horizontal and vertical aspects of an image forming in different planes.

Bandwidth: The range of frequencies over which a given instrument will operate.

Beamsplitter: An optical component which divides a beam into two or more separate beams. A 50:50 beamsplitter produces two beams of approximately equal intensities. A 90:10 beamsplitter transmits approximately 90% of the incident beam and reflects 10% into the second beam.

Beat Frequency: A new frequency formed by the presence of two slightly different frequencies. The beat frequency is equal to the difference in frequency between the original two.

Benton Hologram: Another term for rainbow hologram. Named for its inventor, Steve Benton. A hologram produced by reducing vertical information in order to correct for image dispersion.

Biconcave: A lens which has both faces curving inward. A type of negative lens.

Biconvex: A lens which has both faces curving outward. A type of positive lens

Birefringence: The separation of a beam of light into two beams after it penetrates a doubly refracting object. In monochromatic systems, the two beams interfere, causing undesirable "rings."

Bleach: In holographic processing, a chemical used to change an absorption hologram into a phase hologram to improve efficiency (brightness).

Bragg Diffraction (Bragg's Law): Diffraction which is reinforced by reflection by a series of regularly spaced planes which correspond to a certain wave length and angular orientation. The angle at which this reinforcement occurs is Bragg's angle.

BRH: Bureau of Radiological Health. U.S. government agency responsible for setting laser safety standards.

Brightness: The amount of light perceived.

Cavity: Another name for optical cavity/laser cavity.

Chromatic Aberration: Lens or hologram irregularity due to the shifting of image position for each frequency. If severe enough, the image will appear to blur due to the lack of registration of the colors.

Coherence Length: With respect to laser light, the greatest distance between two components of light (i.e., 2 beams) such that interferometric effects will occur. In holography, the coherence length of the laser will determine the depth over which an object can be recorded.

Coherent Light: Light which is of the same frequency and vibrating in phase. The laser produces coherent light.

Collage: In holography, a technique used to superimpose various spaces on top of one another by overlapping individual holograms or exposures.

Collimated Light: Light which forms a parallel beam and neither converges nor diverges. Also referred to as collimated beam.

Collimator: A device used to produce collimated light by positioning a light source at the focal point of a lens or parabolic mirror. Such a device is called a collimating lens or collimating mirror, respectively.

Color Spread: The area over which a spectrum is dispersed.

Computer-Generated Hologram: A synthetic hologram produced using a computer plotter. The binary structure is produced on a large scale and then photographically reduced into a given medium. The technique allows the production of impossible or nonexistent 3-dimensional forms.

Concave Lens: A lens with an inwardly curving surface which causes light to diverge. See also Negative lens.

Concave Mirror: A mirror with an inwardly curving surface which causes light to converge.

Continuous Wave Laser: A laser which emits a beam which does not vary over time.

Convergence: The optical bending of light rays toward each other, as by a convex lens or concave mirror.

Convex Lens: A lens with an outwardly curving surface which causes light to converge, usually to a focal point.

Copy Hologram: Another term for image plane hologram or any second generation hologram produced from a master hologram. A contact copy is produced by placing the plate in contact with the original.

Copy Plate: Another term for copy hologram. Usually refers to the plate before it is exposed.

Cross Hologram: Another name for the type of holographic stereogram which incorporates the advantages of rainbow holography. Named for Lloyd Cross.

Cross Talk: The phenomenon of spurious images formed by color holograms when an in terference pattern formed by one color also reconstructs an image in another color.

Cylindrical Mirror/Lens: An optical component which cuases light to focus as a slit or line by passing through or reflecting from a surface curved in one dimension.

Denisyuk Hologram: Another name for single beam reflection hologram. Named for its inventor, Y. N. Denisyuk.

Density: The amount of opacity or darkness of a medium.

Depth of Field: The area within which satisfactory resolution of an image can be obtained. Also, in holography, used to describe the area within which any image can be formed, due to the constraints of coherence length.

Developer: A chemical solution which changes the latent image of a photographic image or holographic interference pattern (silver salts) into black metallic silver. The term development usually refers to the degree of effect of the developer or the cause of the amount of density.

Dichromated Gelatin(D.C.G): A light-sensitive emulsion made up of a solution of dichromate com pound, usually ammonium dichromate, in the presence of a gelatin substrate. Exposure results in the crosslinking of gelatin molecules with those of the dichromate compound.

Diffraction: The change in direction of a wave front by encountering an object. Usually refers to the case whereby light is bent by passing through a small aperture.

Diffraction Efficiency: In a hologram, the percentage of incident illumination light diffracted into forming the image. The greater the diffraction efficiency, the brighter the image will appear in a given light.

Diffraction Grating: A holographic diffraction grating is a hologram formed by the interference of two or more beams of pure, undiffused laser light.

Diffuse Reflector: An object that scatters illumination striking it. Most objects are diffuse reflectors.

Distances: In holography, usually refers to the matching of beam path lengths in order to maintain coherence.

Divergence: The bending of light rays away from each other, usually by concave lens or convex mirror, so that the light spreads out. Light will also diverge with a convex lens or concave mirror after it passes through the focal point.

Double Exposure: The formation of two holograms on the same recording medium. Used to cause either overlapping images or two discrete images to appear under different conditions.

Electromagnetic Radiation: Radiation emitted from vibrating, charged particles, all of which travels through space at the speed of light. Visible light is only a small part of the entire electromagnetic spectrum.

Emulsion: The light-sensitive material, including a base, usually of gelatin, which is applied to either a film or glass plate substrate.

Exposure: The act or time of allowing light to impinge upon the emulsion.

Film Plane: The plane at which the recording material is located.

Film Speed: The degree of sensitivity of the film to light to cause a satisfactory exposure. Holographic film speed is usually expressed in ergs/cm.

Fixer: A chemical solution which removes the unexposed silver salts from the emulsion in order to desensitize it and preserve the record stored as metallic silver.

f-number: The ratio of the focal length of a lens or curved mirror to its diameter.

Focal Length: The distance from the center of a lens or curved mirror to a position at which light will converge to a point. Using an element with a positive focal length will cause light to actually focus to a point. Optics with a negative focal length only appear to focus at an imaginary point. The focal point is the place at which the light focuses. The focal plane is a plane through this point.

Focused Image Hologram: A one-step image plane hologram made by using a lens to focus an image directly into the film plane.

Fog: The darkening or exposing of film by inadvertently allowing ambient light to strike it. In holography, a fogged plate reduces fringe contrast, resulting in a less efficient image.

Fourier Transform Hologram: A special hologram formed by a plane object situated in the focal plane of a lens. Also called Fraunhofer Hologram

Frequency: The number of crests of waves that pass a fixed point in a given unit of time.

Fresnel Hologram: Another name for the common hologram. Defined as a hologram formed with an object located close to the recording medium.

Fringe: An individual interference band, made up of one cycle of constructive and destructive interference.

Front Surface Mirror: A mirror with the reflecting surface on the front. Conventional mirrors have their reflecting surfaces on the back of a piece of glass and are not useful for holography as the front surface will produce a "ghost" reflection.

Gabor Hologram: An in-line hologram of the type invented by Dennis Gabor.

Gas Laser: The most common form of laser, which operates by causing the atoms in a gaseous mixture to undergo a population inversion.

Ghost Image: A duplicate image, usually unwanted, which usually is formed as a result of light reflecting from an undesired surface.

Grating (also Diffraction Grating): A device which bends light. A hologram is a special type of grating. Gratings can be made by etching, deposition, acoustically, or holographically.

Ground State: The condition whereby an atom achieves its lowest energy state and greatest stability. Atoms in the ground state are not capable of emitting radiation, but must first be raised to an excited state.

H-1: Another name for master hologram.

Helium-Neon (HeNe): The most common lasing material, which produces a continuous red beam at 632.8 nm.

HOE: Holographic Optical Element. A hologram which may be used to act as a lens, mirror, or some complex optical component.

Hologram: An interference pattern formed as a result of reference light encountering light scattered by an object and stored as such on a light sensitive emulsion.

Holographic Movie: The animation of a 3- dimensional holographic image by presentation of numerous holograms in rapid sequence in much the same way motion picture film operates. Unlike conventional cinema, it is only with extreme difficulty that the image can be projected, and true holographic movies are still very experimental. The term is often used, incorrectly, for holographic stereograms.

Holographic Snapshot: A hologram which merely
replicates an existing object, without any creative input on the part of the holographer.

Holographic Stereogram: A hologram made by filming numerous angles of view of a scene and then storing the frames holographically. Each eye views a different frame, displaced so as to result in the illusion of a stereoscopic image. Also called a multiplex or integral hologram.

Hypo-clear: Chemical solution which removes the fixer. Use of hypo-clear usually improves the stability of the hologram by preventing degradation caused by oxidation of residual fixer.

Image Plane Hologram: A second generation hologram formed by positioning a light sensitive plate in the plane of an image formed by a master hologram.

Immersion Lens: A lens formed by filling a frame with an index-matching liquid to approximate the function of a solid lens, yet at much less cost and with much greater flexibility.

Incandescent Light: Light formed when an electric cunent passes through a resistant metal wire, usually situated in a vacuum bulb.

Incoherent Light: Light which is not in phase with itself. Most light is incoherent.

Index of Refraction: The ratio of the velocity of light in air to the velocity of light in a refractive material for a given wavelength.

Inertial Mass: A large mass which, if at rest, tends to remain at rest

Infrared: That part of the spectrum characterized by wavelengths somewhat longer than those of red light, which are not visible to the eye yet are often perceived as heat Covers the spectrum from about 750 nm. to 1000 micrometers.

In Line Hologram: A Gabor hologram . Made by positioning the object and reference light along the same axis, resulting in a configuration practical only for making holograms of transparencies.

In Phase: The relationship of two waves of the same frequency when they travel through their maximum and minimum values simultaneously and are also polarized identically. Holograms must be made by waves which remain in phase during the course of an exposure.

Interference: The combining of two waves so that their amplitudes add at every point. 'When two coherent waves are so superimposed, the result is either an increase in amplitude (constructive interference) or decrease in amplitude (destructive interference).The result is an n amplitude (destructive interference). The interference pattern records the relative phase relationships between the two waves, thus storing the characteristics of the individual waves. This is how a hologram works.

Interferometer: A device that utilize interference of light to measure changes in systems with extreme accuracy. An interferometer can be used to test the stability of holographic systems.

Ion: An atom which has gained or lost an electron so that it acquires a positive or negative charge.

Ion Laser: A laser within which stimulated emission occurs as a result of energy changes between two levels of an ion. Argon and krypton are the two most common types of ion laser.

Krypton Laser: An ion laser that produces many frequencies which appear over a large part of the spectrum. The most common lines are blue, green, yellow, as well as a very strong red frequency.

Laser: "Light Amplification by Stimulated Emission of Radiation." A laser cavity is filled with a material which, when stimulated with the proper energy, produces a population inversion of excited atoms. The light produced continues to increase in intensity as it oscillates back and forth between two mirrors placed at each end of the cavity. The front mirror is designed to transmit 1 or 2 percent of this amplified light, resulting in a beam of very intense, monochromatic, coherent light.

Latent Image: The image or pattern stored in an emulsion before it is developed into a visible image.

Latent Image Decay: A condition that is common to fine grained silver emulsions, including the types used for holography. The decay occurs if the material is not processed soon after exposure, resulting in a lower density.

Leith, Upatnieks Hologram: Another name for the off-axis hologram, named for its inventors.

Light Meter: Any device used to sense and measure light. Usually used to sense intensity in order to determine exposure.

Line Spacing: The distance between individual interference fringes in a diffraction grating.

Master Hologram: Any hologram which produces an image from which another hologram is made.

Matte Screening: A special type of double exposure whereby two different overlapping yet distinct spaces can be made to appear simultaneously.

Medium: Any substance or material which may be used to convey an idea. In physics, any material through which radiation can travel.

Metal Mounts: Devices to hold optical elements on a vibration isolation table with a metal top.

Metal Table: A vibration isolation system with a metal top.

Mode: The degree to which the beam of a laser is spatially coherent. The tem00 mode is characterized by an even spread of light. The tem01 mode, or donut mode, appears as a ring of light with a dark center.

Moire Pattern: A highly visible type of interference pattern formed when gratings, screens, or regularly spaced patterns are superimposed upon one another.

Monochromatic: Light or other radiation with one single frequency or wavelength. Since no light is perfectly monochromatic, the term is used loosely to describe any light of a single color over a very narrow band of wavelengths.

Motion: The effects of an object or holographic system not remaining rigidly fixed during exposure.

Multichannel Hologram: A hologram formed with two or more separate reference beams or angles.

Multiple Exposure: More than one exposure occurring on the same plate or film.

Multiplex: A type of holographic stereogram.

NAH: A holographic plate without an antihalation backing. Also called an unbacked plate.

Negative Lens: A lens characterized by a concave surface which causes light to diverge. A negative lens has a negative focal length.

Newton's Rings: The series of rings or bands which appear due to interference between two nearly parallel surfaces. These rings often form as a result of light interacting between the front and back surfaces of a holographic plate.

Node: The part of a vibrating wave that is not moving - zero point. An antinode is a point on the wave of maximum displacement from the zero point.

Noise: The effects of undesired light scattered by an emulsion which interfere with the resolution of the image.

Object Beam: The light from the laser which illuminates the object. Also used to describe the entire beam path from the first beamsplitter to the object and then to the plate. In image-plane holography, it is referred to as the master illumination beam .

Off-Axis: The type of hologram invented by Emmet Leith anbd Juris Upatnieks whereby object and reference beams approach the holographic plate at different angles.

On-Axis: Hologram formed with object and reference beams originating along the same axis. Also called an in-line or Gabor hologram.

Open Aperture: A transmission image plane hologram viewable in white light and characterized by both vertical and horizontal parallax and usually a brilliant white image.

Optical Cavity: The space between the two mirrors in a laser. The tube is located within the optical cavity.

Optical Cement: An adhesive used to join optical surfaces. Also useful for laminating a cover glass onto a hologram to protect it. Also known as sealant

Optical Component: An optical device consisting of the optics (lens, mirror, etc.) and a mount used to affix it to a vibration isolation table.

Optics: Those devices which change or manipulate light, including lenses, mirrors, beamsplitters, filters, etc. Also the science of electromagnetic radiation, its effects, and the phenomenon of vision.

Optics Table: Another name for vibration isolation table. The table upon which holograms are made.

Orthoscopic: An image with correct parallax and front to back orientation.

Oscillator: Any device that converts energy into an alternating electromagnetic field, usually of constant period.

Overexposure: Improper exposure resulting from too much light or light reaching the plate or film for too long.

Parabolic Mirror: A mirror with a surface curved in the shape of a parabola. Used as a telescope mirror in astronomy or as a collimating mirror in holography.

Parallax: The difference between two different views of an object, obtained by changing viewing position.

Period: One complete cycle of a wave.

Perspective: The concept of the relationship of various objects in a viewing zone taken from a particular point.

Phase: The position of a wave in space measured at a particular point in time.

Phase Hologram: A means whereby information is stored as a result of phase shifts between two wave fronts.

Phase Shift: The relationship between the phase of one wave front which does not match the phase of another. The result of phase shifts of coherent light is interference.

Photochemistry: Any chemistry related to the action of or upon light sensitive materials.

Photo flo: A chemical wetting agent used to produce an even coating of water over an emulsion during processing to promote even drying.

Photon: The smallest unit or quantum of electromagnetic energy known today.

Photopolymer: A light sensitive plastic which is useful for real-time holography due to almost instantaneous processing.

Photoresist: A chemical substance made insoluable by exposure to light (usually ultraviolet). Although most often used to manufacture microcircuits, photoresist can be used to make holograms.

Pinhole: The small hole used to pass focused light from the objective in a spatial filter.

Plane Hologram: A hologram for which fringes are large with respect to the thickness of the emulsion, so that interference is mostly stored on the surface of the hologram.

Plane Waves: Waves whch propagate as parallel planes, making up a collimated beam.

Plane-Concave: A lens which has one concave surface and one flat surface.

Plane-Convex: A lens which has one convex surface and one flat surface.

Plateholder or Platen: Any device which holds a holographic plate or film in place during exposure.

Polarization: The restriction of light or other radiation to vibration in only one plane.

Population Inversion: A condition whereby more atoms are in the excited state than in the ground state, resulting in the predominance of stimulated emission.

Positive Lens (convex lens): A lens with an outwardly curving surface which causes light to converge.

Processing: The entire chemical sequence, from development to final drying of the hologram.

Pseudocolor: The production of colors in a hologram which are not related to the true colors of the original objects. Usually used in connection with multicolor holograms.

Pseudoscopic: The opposite of orthoscopic. An image that is turned inside out

Pulse Hologram: A hologram produced with the short burst of light from the pulse laser. May be used to make holograms of live subjects.

Pulse Laser: A laser that emits radiation in a wave of short bursts and is inactive between bursts.

Quantum: The smallest amount that the energy of a wave may be divided into.

Rainbow Hologram: A transmission image plane hologram made by restricting illumination of the master vertically to produce a horizontal slit, in order to compensate for chromatic dispersion when the copy is viewed in white light. Vertical parallax is sacrificed, however. The hologram is named for the fact that the image is viewable in each of the colors of the rainbow, which change as one's vertical viewing position changes. Also called a Benton hologram.

Ratios: Used to describe the relationship between the intensities of reference to object light as measured at the position of a holographic plate in a given setup.

Real Image: An image formed by light which actually focuses in space.

Real Time Holography: A technique whereby a holographic image is superimposed over a real object in order to observe in terference fringes generated by minute changes between the two.

Reconstruction Beam: Light directed at the finished hologram from which the object wave front will be recreated.

Recording Material: Any substance which may be used to record the interference pattern of the hologram.

Reference Angle: The angle at which the reference beam strikes the plate, usually measured in degrees from the plate surface.

Reference Beam: The unmodulated, pure laser light directed at the plate to interfere with the object light.

Reflection Hologram: A hologram made by allowing reference and object light to impinge on opposite sides of the plate. The finished hologram is viewed by allowing light to reflect from it to the observer.

Refraction: The bending of light which occurs when it passes from a medium of one refractive index to that of another. In a phase hologram, refraction causes a "phase delay" which corresponds to the original phase diference between the two stored wave fronts.

Refractive Index: Same as index of refraction.

Regular Transmission Holograms: Transmission holograms which are viewed in laser light.

Resolution: The ability of a film or an optical system to distinguish between two closely spaced points. Film resolution is usually expressed in terms of how many closely spaced lines per millimeter the film can record. Holographic films must be capable of high resolving capability since the interference fringes are often extremely small and closely spaced.

Resonance: A large amount of vibration in a system which is caused by a small stimulus with approximately the same period as the natural vibration period of the large system.

Resonant Cavity: Another name for optical cavity or laser cavity.

Sand-Based System: A vibration isolation system which uses a sandbox table as its inertial mass. Also, simply called sandtable .

Sand Mounts: Optic mounts which may be positioned on sand-based optics tables.

Scatter: Unwanted light which interferes with the making of a good quality hologram.

Settling Time: A period of time between the loading of the plate and the exposure in order to allow ambient vibrations time to dampen.

Setup: The configuration of optical components used to produce a given hologram.

Shadowgram: A hologram made by deliberately moving an object during an exposure, or by using an inhe ently unstable object, in order to produce a 3-dimensional "hole" or shadow where the object was once located.

Shutter: The device used to block the laser beam and then allow it to pass unobstructed for the desired exposure time.

Silver halide: The type of recording material which consists of light-sensitive silver particles suspended in gelatin.

Single beam hologram: A hologram made with one beam which acts as both reference and object illumination beam.

Slab Table: An optics table which uses a concrete slab as part of its inertial mass.

Slit Optics: Any optical device which causes light to be propagated into a line. Usually formed by light interacting with a cylindrical surface.

Solid State Laser: A laser which uses a solid material, such as ruby, as its lasing medium.

Spatial Collage: Holographic collage characterized by overlapping of individual dimensional spaces.

Spatial Filtering: The act of "cleaning up" the light of the laser beam by causing it to focus through a tiny aperture. Only the pure light can focus at the desired point, eliminating the effects of dust, optical surface scratches, etc.

Spatial Frequency: Often used with regard to line spacing in diffraction gratings. The spatial frequency is the reciprocal of line spacing, generally expressed in cycles per millimeter. See also resolution.

Speckle: The grainy appearance of an object, or a holographic image, viewed under laser light. It is caused by light reflecting from minute areas of the object and interfering with itself.

Spectral Reflection: Any reflection from a smooth, polished surface, such as a mirror.

Splitbeam: The act of separating a beam of laser light into two components to separately control the action of reference and object illumination.

Squeegee: A device or action used to remove excess water from the emulsion to facilitate drying.

Stability: The requirement for holographic optical systems to remain motionless during an exposure.

Standing Wave: The combination of two waves of the same frequency and amplitude which are traveling in opposite directions. Also used in a general sense to describe any waves which appear motionless due to each new wave replacing the position of the one before.

Stereogram: An image which creates a 3-dimensional illusion by presenting a different view of an object to each eye.

Stimulated Emission: Radiation produced by incoming radiation of the same phase, amplitude, and frequency.

Stop Bath: The chemical bath immediately following the developer which causes the developer to cease action.

Tem00: The lowest mode of a laser, characterized by a beam which is spatially coherent across the diameter of the beam.

Temporal Coherence: Coherence over time. The degree to which waves will remain coherent over time and distance.

Tension Compression Table: An optics table which is strengthened by tension produced along braces made of tightened, threaded, metal rod. Usually a sand-based system. Invented by Lloyd Cross.

Test Strip: A means of visually determining the correct exposure by making a series of individual exposures of varying times on the same plate. The proper time is determined by selecting the strip which yields the brightest or cleanest image.

Theromoplastic Film: A recording material which works due to the effects of electrostatic forces and heat to produce a deformation corresponding to the interference pattern exposed.

360-Degree Hologram: A hologram made by exposing recording material which completely surrounds an object.

Transfer Mirror: A mirror which redirects light from the laser toward the desired working area on the optics table.

Transmission Hologram: Any hologram viewed by passing light through it, toward the viewing side. Transmission holograms are made by allowing both object and reference light to impinge on the same side of the plate.

Transmittance: The proportion of light transmitted by a medium to that which is incident upon it.

Triethanolomine: A chemical used to change the thickness of the emulsion to produce different color playback, usually with reflection holograms.

Ultraviolet: An invisible part of the spectrum characterized by wavelengths somewhat shorter than violet (approx. 100-400 nm.).

Unbacked Plate (NAH): A holographic plate without an antihalation backing. Essential for reflection holograms.

Variable Beamsplitter (VBS): A beamsplitter whereby the ratio of transmitted to reflected beam changes as the beam intercepts the component at different points.

Vibration Isolation: The practice of removing a system from the effects of ambient vibrations which may induce changes, particularly in optical systems. Vibration isolation must be used in making a hologram to prevent the movement of interference fringes during an exposure.

Virtual Image: An image generated by light which does not actually focus in space; yet it appears to do so. The image can be seen, but neither projected nor imaged onto an emulsion.

Volume Hologram: A hologram made up of fringes which are small with respect to the emulsion thickness. The fringes are thus stacked up through the volume of the emulsion. Volume holograms must be illuminated at the original reference angle due to the effects of Bragg diffraction .

Wave Form: The characteristic shape taken on by a wave front.

Wave Front: The surface of a propagating wave which represents all points equidistant from the light source and is characterized by one period of the wave.

Wavelength: The physical distance over which the complete cycle of one wave occurs. Wavelength is inversely proportional to frequency.

White Light Transmission Hologram: Any transmission hologram which can be displayed using ordinary white light.

YAG Laser: A solid state laser using Yttrium Aluminum Garnet as the lasing material.

Zone Plate: A pattern consisting of a central spot surrounded by concentric zones, alternatingly opaque and transparent, the total area of each zone being equal. It represents the real image of a point produced by diffraction.

INDEX

DID YOU BORROW THIS COPY?

If so, now is the time to order your own personal copy of the **Holography Marketplace 1989**. This international directory for the holography industry is the first and only resource of its kind. You will refer to the HMP day after day, so don't you want one of your own?

☐ **YES !** I want my own personal copy of the 1989 Holography Marketplace.

Price:
within **USA** (US)$37.90 includes UPS shipping

FOREIGN (US)$47.00 includes Airmail shipping overseas.

☐ Check enclosed

☐ Charge my credit card:

☐ MASTERCARD ☐ AMERICAN EXPRESS ☐ VISA

Card Number #: _____

Expiration Date: _____

Signature: _____

Address all correspondence to:

Holography Marketplace
Ross Books
P.O. Box 4340
Berkeley, CA 94704
USA

To order: **CALL TOLL FREE IN THE USA: (1) (800) 367 0930**
or **FAX your order:(1) (415) 841 2695**

NAME _____

TITLE _____

COMPANY _____

ADDRESS _____

CITY/STATE(PROVINCE)/ZIP CODE

_____

_____

PHONE: ( ) _____

ISBN #: 0-89496-047-4
size: 8 1/2" x 11"
pages: 184
binding: softcover
Price: $35.00

www.ingramcontent.com/pod-product-compliance
Lightning Source LLC
Chambersburg PA
CBHW051346200326
41521CB00014B/2491